Manifolds III - Developments and Applications

Edited by Paul Bracken

Published in London, United Kingdom

Manifolds III - Developments and Applications
http://dx.doi.org/10.5772/intechopen.100720
Edited by Paul Bracken

Contributors
Thales Fernando Vilamaior Paiva, Fusun Nurcan, Paul Bracken, Simon Davis

First published in London, United Kingdom, 2022 by IntechOpen
IntechOpen is the global imprint of INTECHOPEN LIMITED, registered in England and Wales, registration number: 11086078, 5 Princes Gate Court, London, SW7 2QJ, United Kingdom

British Library Cataloguing-in-Publication Data
A catalogue record for this book is available from the British Library

Additional hard and PDF copies can be obtained from orders@intechopen.com

Manifolds III - Developments and Applications
Edited by Paul Bracken
p. cm.
Print ISBN 978-1-80356-230-8
Online ISBN 978-1-80356-231-5
eBook (PDF) ISBN 978-1-80356-232-2

Meet the editor

Paul Bracken is a professor in the Department of Mathematics, University of Texas, Edinburg. He received a BSc from the University of Toronto and a Ph.D. from the University of Waterloo, Canada. His research interests include applications of differential geometry and partial differential equations to problems in mathematical physics, as well as problems in quantum mechanics and quantum field theory. He has published more than 180 research papers in journals and books and presented talks at numerous meetings and conferences. He is also a member of many societies, including the American Mathematical Society (AMS), Canadian Mathematical Society (CMS), and American Physical Society (APS).

Contents

Preface

The study of manifolds, including their properties and applications, is a fundamental part of modern physics as well as other domains such as group theory and topology. This book includes five chapters that discuss the mathematics of manifolds and some foundational issues related to physics, as manifolds have many applications to problems in physics.

This book has been assembled out of the hard work of an international group of invited authors. It is a pleasure to thank them for their efforts and scientific contributions. I am grateful to acknowledge with many thanks the staff at IntechOpen, especially Ana Javor for her assistance throughout the publication process.

Paul Bracken
Professor,
Department of Mathematics,
University of Texas,
Edinburg, TX, USA

Section 1

Manifolds - Perspectives

Chapter 1

Free Actions of Compact Lie Groups on Manifolds

Thales Fernando Vilamaior Paiva

Abstract

If a compact Lie group G acts freely on a manifold X, the resulting orbit space X/G is itself a manifold. This text is concerned with the existence of such actions as well as the cohomological classification of the respective orbit spaces by using some known tools of equivariant cohomology theory and spectral sequences.

Keywords: free actions, manifolds, orbit spaces, cohomology, spectral sequences

1. Introduction

When a topological group G acts on a manifold X, we can define the orbit space X/G, that does not necessarily have the structure of a manifold. However, when G is a compact Lie group and we impose the condition that the action be free, which means that the isotropy subgroup G_x contains only the trivial element 1 of G, for any $x \in X$, then we can construct on X/G a manifold structure.

The general situation above can be illustrated by the construction of the projective spaces kP^n, for $k = \mathbb{R}, \mathbb{C}$ or $\mathbb{H}$, that are orbit spaces of certain free actions of the groups $\mathbb{Z}_2$, S^1, and S^3 on spheres S^n, S^{2n+1}, and S^{4n+3}, respectively. Such spaces, as we know, appear in different contexts and for this reason there is an interest in obtaining certain algebraic and geometric invariants that characterize them, such as their homotopic and cohomological classification.

Thus, given X and G, we can naturally consider the general problem of classifying the space of orbits X/G, through the existence or not of a free action of G on X, which is a typical transformation group problem associated with this data.

The interest in this type of problem, particularly when G is a finite group, has become greater since the publication of work [1], by H. Hopf in 1926, in which one formalizes the purpose of classification of all manifolds whose universal covering is homeomorphic to a sphere S^n. This problem, as we know, is equivalent to the classification of all finite groups that can act freely on S^n.

However, we realize that to get a homotopic classification of such spaces can become extremely complicated, even when the space X already has a known classification. For example, let S^n be the $n-$sphere seen as the one-point compactification of euclidean space $\mathbb{R}^n$, which does not have a complete classification of its homotopy groups. As a result, instead of a homotopic classification, we can consider a cohomological classification of these orbit spaces.

In this direction, we realize that the difficulty in computing the cohomology of the orbit space X/G, by direct methods, becomes evident when X has nontrivial cohomology on several levels. On the other hand, many such results have been obtained by using some tools of equivariant cohomology theory. This is due to the fact that as long as G is a compact Lie group acting freely on a space X, there is a homotopy equivalence between the orbit space X/G and the Borel space X_G, so that we can use the so-called Leray-Serre spectral sequence associated with the Borel fibration:

$$X \hookrightarrow X_G \longrightarrow B_G, \tag{1}$$

where B_G denotes the classifying space for group G, to investigate the cohomology ring $H^*(X_G; R) \cong H^*(X/G; R)$.

In this text, we briefly deal with these tools and then we present some applications regarding the existence of free actions of certain compact Lie groups in some classes of smooth manifolds, as well as the cohomological classification of the respective orbit spaces.

2. Preliminary concepts

2.1 Group actions and classifying spaces

Let $\mu : G \times X \to X$ be an action of a topological group G on a topological space X, i. e. μ is a continuous map such that

$$\mu(g, \mu(h, x)) = \mu(gh, x), \tag{2}$$

$$\mu(e, x) = x, \tag{3}$$

for any $g,h \in G$ and for any $x \in X$, where e indicates the neutral element of G. In this case, we say that G acts on X and X is a $G-$space.

As it is usual, we denote by $\mu(g, x) = g(x)$ or simply $\mu(g, x) = gx$ to indicate the action of the element g of G on $x \in X$.

For each $x \in X$, the subspace $G(x) = \{gx; g \in G\}$ is called the orbit of the element x. It is a simple task to show that for any two orbits $G(x)$ and $G(y)$, then $G(x) \cap G(y) = \emptyset$ or $G(x) = G(y)$. Therefore, we can define the orbit space:

$$X/G = \{G(x); x \in X\}, \tag{4}$$

which is provided with the quotient topology induced by the natural map $q : X \to X/G$, given by $q(x) = G(x)$, which is called orbit map.

Example 2.1.1. Any group G acts on itself by multiplication. Precisely, we can define $\mu : G \times G \to G$ by $\mu(g, h) = gh$.

An action μ of G on X induces a group homomorphism $\Gamma_\mu : G \to \text{Homeo}(X)$, such that, for each $g \in G$, we define $\Gamma_\mu(g) = L_g$, where

$$L_g : X \to X, L_g(x) = gx. \tag{5}$$

The action μ is called effective when the kernel of the homomorphism Γ_μ contains only the trivial element $e \in G$ and is called trivial when $\ker\Gamma_\mu = G$.

For each $x \in X$, we call the isotropy subgroup at x the following subgroup of G:

$$G_x = \{g \in G; gx = x\}. \tag{6}$$

When $G_x = \{e\}$, for any point $x \in X$ then the action is called free, and X is said to be a free $G-$space. The set $X^G = \{x \in X; gx = x\}$ is called the fixed point set of the action.

Remark 2.1.1. If X is a haursdorff space and G is a compact space, it is well known that any action $\mu : G \times X \to X$ is a closed map, according to Theorem 1.2 of [2]. Furthermore, in this case, the subspace $\mu(G \times A) \subseteq X$ is closed (resp. compact) if A is closed (resp. compact).

Let X and Y be $G-$spaces. If the map $f : X \to Y$ is equivariant, i. e. $f(gx) = gf(x)$, for any $g \in G$ and any $x \in X$, then we can define the map:

$$\bar{f} : X/G \to Y/G, \ \bar{f}(G(x)) = G(f(x)). \tag{7}$$

Remark 2.1.2. Even though it is possible to investigate actions in arbitrary topological spaces, we are interested in observing certain structures both in X and in the orbit space X/G, so that we will assume, from now on, that X is a manifold (smooth or not) and G is a Lie group.

We recall that a Lie group G is a topological group that is also a (real) finite-dimensional smooth manifold, in which the multiplication operation $(g_1, g_2) \mapsto g_1 g_2$ and the inversion map $g \mapsto g^{-1}$ are smooth.

Example 2.1.2. The matrix groups $GL(n, \mathbb{R})$, $GL(n, \mathbb{C})$, of the invertible $n \times n-$matrices with entries in $\mathbb{R}$ or $\mathbb{C}$, respectively, are standard examples of Lie groups, along with respective subgroups (special linear groups) $SL(n, \mathbb{R})$ and $SL(n, \mathbb{C})$.

Example 2.1.3. Let $O(n) \subset GL(n, \mathbb{R})$ be the subgroup of the orthogonal matrices, i. e. those in which $AA^t = Id$ and let $U(n)$ be the subgroup of $SL(n, \mathbb{C})$ of the unitary matrices $\overline{A}A^t = Id$. We can define the special orthogonal group by $SO(n) = O(n) \cap SL(n, \mathbb{R})$ and the special unitary group by $SU(n) = U(n) \cap SL(n, \mathbb{C})$.

We know that $SU(2)$ is isomorphic to S^3 identified as the subgroup of the unitary quaternions. Also we know that the isomorphic groups $U(1)$ and $SO(2)$ are isomorphic to the circle group S^1.

For each integer m, let $\mathbb{Z}_m = \mathbb{Z}/m$ be the group of the integer modulo m, which can be identified with the subgroup of S^1 of all $m-$th roots of unity. In particular, we have the following chain of Lie (sub)groups:

$$\mathbb{Z}_m \subset S^1 \cong U(1) \cong SO(2) \subset SU(2) \cong S^3. \tag{8}$$

Example 2.1.4. (Free actions on spheres) Let X be the $n-$sphere $S^n \subset \mathbb{R}^{n+1}$ and G be the finite group $\mathbb{Z}_2$. Then, G acts freely on X by the antipodal map $\mu(1, x) = A(x) = -x$, and in this case we have $X/G = \mathbb{R}P^n$ the real projective space.

For $X = S^{2n-1} \subset \mathbb{R}^{2n} \cong \mathbb{C}^n$ and $G = S^1$ seen as a subgroup of the complex plane, we can consider the free action induced by complex multiplication:

$$\mu : G \times X \to X, \mu(z, (z_1, \cdots, z_n)) = (zz_1, \cdots, zz_n), \tag{9}$$

and it follows that $X/G = \mathbb{C}P^{n-1}$ is the complex projective space.

Let $X = S^{4n-1} \subset \mathbb{R}^{4n} \cong \mathbb{H}^n$ and $G = S^3$ identified with the group of the unitary quaternions $SU(2)$, where $\mathbb{H}$ denotes the quaternion algebra:

$$\mathbb{H} = \left\{ \begin{bmatrix} \alpha & -\overline{\beta} \\ \beta & \overline{\alpha} \end{bmatrix}; \alpha, \beta \in \mathbb{C} \right\}. \tag{10}$$

Similar to the previous case, we can define the free action $\mu : G \times X \to X$, induced by the multiplication, such that $X/G = \mathbb{H}P^{n-1}$ the quaternionic projective space.

Example 2.1.5. (Free involutions on projective spaces) An involution[1] on a space X is a continuous action of the group $\mathbb{Z}_2$ on X. Let $[x_1, x_2, \cdots, x_{2n-1}, x_{2n}] \in \mathbb{R}P^{2n-1} = S^{2n-1}/\mathbb{Z}_2$ be an arbitrary element. Its easy to see that the map:

$$T([x_1,\ x_2,\ \cdots,\ x_{2n-1},\ x_{2n}]) = [-x_2, x_1, \cdots, -x_{2n}, x_{2n-1}] \tag{11}$$

defines a free involution on $\mathbb{R}P^{2n-1}$.

Similarly, for $[z_1, z_2, \cdots, z_m, z_{m+1}]$ an arbitrary element in $\mathbb{C}P^m = S^{2m+1}/S^1$, if $m > 1$ is odd then we can define the free involution:

$$S([z_1,\ z_2,\ \cdots,\ z_m,\ z_{m+1}]) = [-\overline{z}_2, \overline{z}_1, \cdots, -\overline{z}_{m+1}, \overline{z}_m] \tag{12}$$

$$\mathbb{C}P^m. \tag{13}$$

When a group G acts on a manifold X, in general, we can consider the orbit space X/G with no other additional structure. However, when the action is free and proper, then this orbit space can be seen as a manifold, according to the Quotient Manifold Theorem:

Theorem 2.1.1. (Quotient Manifold Theorem) [4]. Let G be a compact Lie group acting freely (and smoothly) on a smooth manifold X. If the action is free and proper, then X/G is also a smooth manifold of dimension $\dim(X) - \dim(G)$, such that the quotient map $X \to X/G$ is a principal $G-$bundle and a smooth submersion.

Remark 2.1.3. As an immediate consequence of the previous theorem, for every cohomology functor H^* we have $H^j(X/G; R) = \{0\}$, for all $j > \dim X - \dim G$, for any commutative ring with unity R.

Recall that given any compact Lie group G we can construct the universal $G-$bundle $p_G : E_G \to B_G$, with fiber space G, where the total space E_G is the $G-$space defined as the join operation[2] of infinite copies of G, and the base is the quotient space (by diagonal action) $B_G = E_G/G$, which is called the classifying space for G and p_G is the projection.

Example 2.1.6. (Classifying spaces for $\mathbb{Z}_2$,S^1 and S^3) For $G = \mathbb{Z}_2$, we can see that $B_G = E_G/G \cong \mathbb{R}P^\infty$. Consequently, the mod 2 cohomology of the classifying space B_G is given by $H^*(B_G; \mathbb{Z}_2) = \mathbb{Z}_2[t]$, where $\deg t = 1$. Regarding to $G = S^1$, since $B_G = E_G/G \cong \mathbb{C}P^\infty$, then $\pi_1(B_G) = 1$ and the mod 2 cohomology is give by $H^*(B_G; \mathbb{Z}_2) = \mathbb{Z}_2[\tau]$, where $\deg\tau = 2$. With respect to the group $G = S^3$, it follows that $B_G = E_G/G \cong \mathbb{H}P^\infty$. Since[3] $\pi_i(B_G) \cong \pi_{i-1}(G)$, therefore $\pi_1(B_G) \cong 1$ and the mod 2 cohomology of B_G is given by $H^*(B_G; \mathbb{Z}_2) = \mathbb{Z}_2[\tau]$, where $\deg\tau = 4$.

---

[1] Formally, an involution is a map $T : X \to X$ such tha $T^2 = 1$. For an extensive treatment of involutions on manifolds, see [3].

[2] For more details on Milnor's construction of classifying spaces, see [5] and Section 4.11 of [6].

[3] For more details, see Corollary 8.13 of [7].

2.2 The Leray-Serre spectral sequence of a Borel fibration

Let X be a free $G-$space, $q : X \to X/G$ the orbit map, and $\varpi : E_G \to B_G$ the universal $G-$bundle. The group G acts freely on the product $X \times E_G$, by the diagonal action $g(x, y) = (gx, gy)$, where do we get

$$\rho : X \times E_G \to (X \times E_G)/G = X_G \tag{14}$$

the respective orbit map. The quotient space X_G is also know as the Borel space.

Since the projections $proj_1 : X \times E_G \to X$ and $proj_2 : X \times E_G \to E_G$ are $G-$equivariant, they induce the fibrations π and p, respectively, according to the diagram below:

$$\begin{array}{ccccc} E_G & \xleftarrow{proj_2} & X \times E_G & \xrightarrow{proj_1} & X \\ \downarrow{\scriptstyle \varpi} & & \downarrow{\scriptstyle \rho} & & \downarrow{\scriptstyle q} \\ B_G & \xleftarrow{p} & X_G & \xrightarrow{\pi} & X/G \end{array} \tag{15}$$

where π is called the Borel fibration with fiber X, and p is a principal $G-$bundle.

Moreover, under the above hypothesis, the fiber E_G of p is contractible and therefore p is a homotopy equivalence[4], which induces a natural isomorphism $p^* : H^*(X/G; R) \to H^*(X_G; R)$, for any commutative ring with unit R.

By Theorem 5.2 of [9], there is a first quadrant cohomological spectral sequence $\{E_r^{*,*}, d_r\}$ converging to $H^*(X_G; R) \cong H^*(X/G; R)$, as an algebra, such that the E_2-page $E_2^{p,q}$, is isomorphic to

$$E_2^{p,q} \cong H^p(B_G; \mathscr{H}^q(X; R)), \tag{16}$$

where the symbol $\mathscr{H}^q(X; R)$ indicates a system of local coefficients twisted by the action of the fundamental group $\pi_1(B_G)$ on the cohomology ring of X.

When $\pi_1(B_G)$ acts trivially on $H^*(X; R)$, the system of local coefficients $\mathscr{H}^q(X; R)$ is simple and, according to Proposition 5.6 of [9], the E_2-page as in (13) takes the form:

$$E_2^{p,q} \cong H^p(B_G; R) \otimes_R H^q(X; R), \tag{17}$$

what happens, in particular, when $\pi_1(B_G) = 1$.

Moreover, by Theorem 5.9 of [9], the homomorphisms

$$H^q(B_G; R) = E_2^{q,0} \twoheadrightarrow \cdots \twoheadrightarrow E_q^{q,0} \twoheadrightarrow E_{q+1}^{q,0} = E_\infty^{q,0} \subseteq H^q(X_G; R) \tag{18}$$

and

$$H^q(X_G; R) \twoheadrightarrow E_\infty^{0,q} = E_{q+1}^{0,q} \subseteq E_q^{0,q} \subseteq \cdots \subseteq E_2^{0,q} = H^q(X; R) \tag{19}$$

coincide with the homomorphisms $\pi^* : H^q(B_G; R) \to H^q(X_G; R)$ and $i^* : H^q(X_G; R) \to H^q(X; R)$, respectively.

---

[4] According to ([8], p. 180.)

3. Some applications

Supposing that there is a free action of a group G on X, there is an interest in classifying the orbit space X/G, in the same way as it happens on the real, complex, and quaternionic projective spaces.

In order to do this, we will cite some results that use these tools presented in the previous sections to obtain cohomological classifications for certain known spaces.

However, we observe that in general way, the computation of the cohomology of the orbit space X/G can be a difficult task when the space X is not a sphere or, more generally, when X has nontrivial cohomology $H^j(X;R) \neq \{0\}$ on several levels $0 < j < \dim X$

Remark 3.1. Since all these results use only the cohomological struture of space X, we observe that the same conclusions can be obtained replacing X by any finitistic space that has same Čech cohomology algebra. Recall that a finitistic space is a paracompact Hausdorff space whose every open covering has a finite-dimensional open refinement, where the dimension of a covering is one less than the maximum number of members of the covering which intersect nontrivially. It is known [10, 11] that if G is a compact Lie group acting continuously on X, then X is finitistic if and only if the orbit space X/G is finitistic. Therefore, we can consider the problem of cohomology classification of the orbit spaces up to finitistic spaces of isomorphic cohomology to initial space X.

3.1 Free actions on spheres and projective spaces

In 1926, Hopf posed the general problem of classifying all groups that present freely in S^n. Posteriorly, in 1957, J. Milnor provided some answers for this problem by showing, among other things, that the symmetric group S_3 cannot act freely on S^n.

Even considering this classification only on the category of compact Lie groups, this problem still does not have a complete solution for a arbitrary sphere S^n. However, if n is even, the only finite group that acts freely on S^n is the group $\mathbb{Z}_2$.

In fact, if G acts freely on $X = S^{2k}$, then the quotient map $X \to X/G$ is a covering projection; therefore, with $\chi(\cdot)$ the Euler characteristic, it follows that

$$2 = \chi(S^{2k}) = |G| \cdot \chi(S^{2k}/G), \tag{20}$$

which implies that $|G| = 1$ or $|G| = 2$. Since the action is free, the only possibility is $|G| = 2$ and then $G = \mathbb{Z}_2$. Furthermore, the resulting orbit space $S^n/\mathbb{Z}_2$ has the same homotopy type of real projective space $\mathbb{R}P^n$.

For n odd, the Section 3.8 of [2] contains a compilation of results related to the existence of free G−actions on spheres S^n. In particular, with respect to groups of positive dimension, the Theorem 8.5 of [2] states that a group G that act freely on S^n must be isomorphic to S^3, S^1, or $N(S^1)$.

Suppose that $G = S^1$ and let $X = S^n$ and $X_G \to B_G$ the Borel fibration.

For $\{E_r^{*,*}, d_r\}$ the associated Leray-Serre spectral sequence, we have

$$E_2^{p,q} = H^p(B_G;\mathbb{Z}_2) \otimes H^q(X;\mathbb{Z}_2) = \begin{cases} \mathbb{Z}_2 \otimes \mathbb{Z}_2, \text{ if } p \text{ is odd and } q = 0, n \\ \{0\}, \text{ otherwise.} \end{cases}$$

Since this sequence converges to $H^*(X/G; \mathbb{Z}_2)$, it follows that it cannot collapse on E_2–page, which means that there is a nontrivial differential $d_{2k} : E_{2k}^{0,n} \to E_{2k}^{2k,0}$, for $k \in \mathbb{Z}$ such that $n = 2k - 1$.

Therefore, the sequence collapse on E_{2k+1}–page is $E_\infty = E_{2k+1}$, whose only nonzero row is $E_\infty^{\text{even},0}$ and the total complex is isomorphic to the graded cohomology ring $H^*(\mathbb{C}P^{n-1}; \mathbb{Z}_2)$.

Proceeding in a similar way, we can show that $H^*(S^{4n-3}/S^3; \mathbb{Z}_2) = H^*(\mathbb{H}P^{n-1}; \mathbb{Z}_2)$.

In order to generalize this type of problem, we can consider X a product of spheres. For example, L. W. Cusick [12] showed that if a finite group G acts freely on a product of spheres of even dimensions, $X = S^{2n_1} \times \cdots \times S^{2n_k}$, then then G must be isomorphic to a group of the type $\mathbb{Z}_2^r$, for some $r \leq k$.

Concerning on free actions of a finite group $\mathbb{Z}_p$, p prime, and the circle group S^1 on a product of spheres $S^m \times S^n$, Dotzel et al. [13] showed the following classification results according to Theorems 3.1.1, 3.1.2, and 3.1.3.

Theorem 3.1.1. Let $\mathbb{Z}_p$, p an odd prime, act freely on $X = S^m \times S^n$, $0 < m \leq n$. Then, $H^*(X/\mathbb{Z}_p; \mathbb{Z}_p)$ is isomorphic to $\mathbb{Z}_p[x, y, z]/\phi(x, y, z)$, as a graded commutative algebra, where $\phi(x, y, z)$ is one of the following ideals:

i. $(x^2, y^{(m+1)/2}, z^2)$ m odd, $\deg x = 1$, $y = \beta(x)$ the Bockstein cohomology operation and $\deg z = n$;

ii. $(x^2, y^{(m+n+1)/2}, y^{(n-m+1)/2}z - ay^{(n+1)/2}, z^2 - by^m)$, m even, n odd, $\deg x = 1$, $y = \beta(x)$, $\deg z = m$, $a,b \in \mathbb{Z}_p$, and $a = 0$ necessarily when $n < 2m$;

iii. $(x^2, y^{(n+1)/2}, z^2 - by^m)$, n odd, $\deg x = 1$, $y = \beta(x)$, $\deg z = m$, $b \in \mathbb{Z}_p$, $b \neq 0$ only when m is even and $2m < n$.

Theorem 3.1.20. Let $\mathbb{Z}_2$ act freely on $X = S^m \times S^n$, $0 < m \leq n$. Then, $H^*(X/\mathbb{Z}_2; \mathbb{Z}_2)$ is isomorphic to $\mathbb{Z}_p[y, z]/\psi(y, z)$, as a graded commutative algebra, where $\psi(y, z)$ is one of the following ideals:

i. (y^{m+2}, z^2), $\deg y = 1$ and $\deg z = n$;

ii. $(y^{m+n+1}, y^{n-m+1}, z, z^2 - ay^m z - by^{2m})$, $\deg y = 1$, $\deg z = m$, $a,b \in \mathbb{Z}_2$, and $a = 0$ necessarily when $n < 2m$;

iii. $(y^{n+1}, z^2 - ay^m z - by^{2m})$, $\deg y = 1$, $\deg z = m$, $a,b \in \mathbb{Z}_2$, and $b = 0$ necessarirly when $m = n$ or $n < 2m$.

Theorem 3.1.3. Let $G = S^1$ act freely on $X = S^m \times S^n$, $0 < m \leq n$. Then, $H^*(X/G; \mathbb{Q})$ is isomorphic to $\mathbb{Q}[y, z]/\psi(y, z)$, as a graded commutative algebra, where $\psi(y, z)$ is one of the following ideals:

i. $(y^{(m+1)/2}, z^2)$, m odd, $\deg y = 2$ and $\deg z = n$;

ii. $(y^{(m+n+1)/2}, zy^{(n-m+1)/2} - ay^{(n+1)/2}, z^2 - by^m)$, m even, n odd, $\deg y = 2$, $\deg z = m$, and $a = 0$ necessarily when $n < 2m$;

iii. $\left(y^{(n+1)/2}, z^2 - by^m\right)$, n odd, $\deg y = 2$, $\deg z = m$, and $b \neq 0$ only when m is even and $2m < n$.

Using the same techniques, it is shown in [14] similar results regarding the action of groups S^1 and S^3 on the product of spheres, considering both rational and mod 2 coefficients.

Theorem 3.1.4. The group S^3 cannot act freely on a $n-$torus $X = \left(S^1\right)^n$.

Proof. Let X be the $n-$torus $\left(S^1\right)^n$ and suppose that $G = S^3$ act freely on X, with $n \geq 3$. Let $x_1, \cdots, x_n \in H^1(X, \mathbb{Z}_2)$ be the generators. By Quotient Manifold Theorem, the spectral sequence $\left\{E_r^{*,*}, d_r\right\}$ associated with the Borel fibration $X_G \to B_G$ does not collapse on the E_2-term. Therefore, there must exist some nontrivial differential $d_r^{p,q}$, for a certain $r \geq 2$, such that

$$E_r^{p,q} \cong E_{r-1}^{p,q} \cong \cdots \cong E_2^{p,q} = H^p(B_G; \mathbb{Z}_2) \otimes H^q(X; \mathbb{Z}_2),$$

and it is clear that this is only possible when $r \geq 4k$, for some $k \in \mathbb{N}$.

Let us suppose that $r = 4$ and let $y = x_{i_1}x_{i_2}x_{i_3} \in H^3(X; \mathbb{Z}_2)$ be an element for which $d_4^{0,3}(1 \otimes y) = \tau \otimes 1$. By dimensional reasons, $d_4^{0,1}(1 \otimes x_i) = 0$ for all $1 \leq i \leq n$; therefore, it follows that

$$\tau \otimes 1 = d_4^{0,3}(1 \otimes y) = (1 \otimes x_{i_1})(1 \otimes x_{i_2})d_4^{0,1}(1 \otimes x_{i_3}) = 0,$$

for τ the generator of $H^*(B_G; \mathbb{Z}_2)$, which is a contradiction. Since this argument works for any $r \geq 4$ and for any $y \in H^j(X; \mathbb{Z}_2)$, $j \geq 3$, it follows that G cannot act freely on X. $\square$

Let $p \geq 2$ be a positive integer and $q_1, \cdots, q_m$ be integers coprime to p, where $m \geq 1$. Then the action of $\mathbb{Z}_p$ on $S^{2m-1} \subset \mathbb{C}^m$ defined by:

$$\left(e^{2\pi i q_1/p}, \cdots, e^{2\pi i q_m/p}\right) * (z_1, \cdots, z_m) = \left(e^{2\pi i q_1/p} z_1, \cdots, e^{2\pi i q_m/p} z_m\right)$$

is itself free. Therefore, the resulting orbit space is a compact Hausdorff orientable manifold of dimension $2m - 1$, which is called lens space and it is denoted by:

$$S^{2m-1}/\mathbb{Z}_p = L_p^{2m-1}(q_1, \cdots, q_m) = L_p^{2m-1}(q).$$

Theorem 31.5. [15] Let $G = \mathbb{Z}_2$ act freely on $X = L_p^{2m-1}(q)$. Then, $H^*(X/G; \mathbb{Z}_2)$ is isomorphic to one of the following graded commutative algebras:

i. $\mathbb{Z}_2[x]/\langle x^{2m}\rangle$, where $\deg x = 1$.

ii. $\mathbb{Z}_2[x, y]/\langle x^2, y^m\rangle$, where $\deg x = 1$ and $\deg y = 2$.

iii. $\mathbb{Z}_2[x, y, z]/\langle x^3, y^2, z^{m/2}\rangle$, where $\deg x = \deg y = 1$, $\deg z = 4$, and m is even.

iv. $\mathbb{Z}_2[x, y, z]/\langle x^4, y^2, z^{m/2}, x^2y\rangle$, where $\deg x = \deg y = 1$, $\deg z = 4$, and m is even.

v. $\mathbb{Z}_2[x, y, w, z]/\langle x^5, y^2, w^2, z^{m/4}, x^2y, wy\rangle$, where $\deg x = \deg y = 1$, $\deg w = 3$, $\deg z = 8$, and $4|m$.

Related to actions of $\mathbb{Z}_2$ on the product of projective spaces, both real and complex, we can mention the work [16], which provides a list of possible cohomology algebras for the respective orbit spaces. In [17], the authors showed that the group $G = S^3$ cannot act freely on the real projective space of any dimension.

Theorem 3.1.6. The group $G = S^3$ cannot act freely on $X = \mathbb{C}P^n, \mathbb{H}P^n$, for any $n > 0$.

Proof. Let us suppose that the group $G = S^3$ acts freely on $X = \mathbb{C}P^n$. Then, the spectral sequence $\{E_r^{*,*}, d_r\}$ associated with the Borel fibration $X \rightarrowtail X_G \to B_G$, which has the E_2−term given by $E_2^{p,q} = H^p(B_G; \mathbb{Z}_2) \otimes H^q(X; \mathbb{Z}_2)$, converges to $H^*(X_G; \mathbb{Z}_2) \cong H^*(X/G; \mathbb{Z}_2)$, as an algebra. By the cohomology structures of $B_G \cong \mathbb{H}P^\infty$ and $X = \mathbb{C}P^n$, it follows that

$$E_2^{p,q} \cong \begin{cases} \mathbb{Z}_2, & \text{if } p = 4i \text{ and } q = 2j, \text{ for all } i, j \geq 0, \\ \{0\}, & \text{otherwise.} \end{cases}$$

Therefore, a differential $d_r^{p,q} : E_r^{p,q} \to E_r^{p+r,q+1-r}$ with bidegree $(r, 1-r)$, is nontrivial only if $p = 4i$ and $q = 2j \leq 2n$, for some positive integers i and j. In this case, we have the following equality involving the bidegrees: $(4i + r, 2j + 1 - r) = (4k, 2l)$, for certain integers $k,l > 0$, that is, these numbers must satisfy the linear system:

$$\begin{cases} 4i + r = 4k, \\ 2j + 1 - r = 2l, \end{cases}$$

that clearly has no integer solution; therefore, we conclude that all differentials $d_r^{*,*}$ are trivial, for all $r \geq 2$. This implies that the sequence collapses on its $E_r \cong E_2$−term and contradicts the Quotient Manifold Theorem.

Similarly, let us suppose that the group S^1 acts freely on $X = \mathbb{H}P^n$, and let us consider $\{E_r^{*,*}, d_r\}$ the spectral sequence associated with the Borel fibration $X_{S^1} \to B_{S^1}$, whose E_2−term is given by $E_2^{p,q} \cong H^p(B_{S^1}; \mathbb{Z}_2) \otimes H^q(X; \mathbb{Z}_2)$.

Let t be the generator of $H^*(\mathbb{C}P^\infty; \mathbb{Z}_2) \cong H^*(B_{S^1}; \mathbb{Z}_2)$ and τ be the generator of $H^*(\mathbb{H}P^n; \mathbb{Z}_2)$. Then,

$$E_2^{p,q} \cong \begin{cases} \mathbb{Z}_2, & \text{if } p = 2i \text{ and } q = 4j,\ i, j \geq 0, \\ \{0\}, & \text{otherwise.} \end{cases}$$

By Quotient Manifold Theorem, the spectral sequence does not collapse on it E_2−term; therefore, there must exist some nontrivial differential $d_r^{*,*}$. If $r \geq 2$ is the smallest integer for which this happens, so that

$$E_r^{p,q} \cong E_{r-1}^{p,q} \cong \cdots \cong E_2^{p,q},$$

for all $p,q \geq 0$, we see that this is only possible when the integers r,i,j, and k (which are obtained from the equality between the bidegrees involved) satisfy the linear system:

$$\begin{cases} r = 2i, \\ 4j + 1 - r = 4k. \end{cases}$$

But this system has no integer solution; therefore, the group S^1 cannot act freely on X. Since S^1 is a subgroup of S^3, then X does not admit any free action of S^3. □

3.2 Free actions on spaces of type (a, b)

Let X be a finite CW complex. We say that X is a space of type (a, b), characterized by an integer $n > 0$, if

$$H^j(X;\mathbb{Z}) = \begin{cases} \mathbb{Z}, & \text{if } j = 0, n, 2n, 3n, \\ \{0\}, & \text{otherwise,} \end{cases} \tag{21}$$

whose generators $u_i \in H^{in}(X;\mathbb{Z})$ satisfy the relations $au_2 = u_1^2$ and $bu_3 = u_1u_2$, for certain integers a and b. By Universal Coefficient Theorem, the mod 2 cohomology of X is given by $H^{in}(X;\mathbb{Z}_2) \cong H^{in}(X;\mathbb{Z}) \otimes \mathbb{Z}_2 \cong \mathbb{Z}_2$, for $n = 0,1,2,3$, and the relations come to depend only on the parity of the numbers a and b. In this case, we will use the same symbols to denote the generators, i. e.

$$\mathbb{Z}_2 \cong \langle u_i \rangle \cong H^{in}(X) = H^{in}(X;\mathbb{Z}_2). \tag{22}$$

Example 3.2.2. The spaces of type (a, b) were first studied by James [18] and Toda [19]. Note that we can construct examples of these spaces by considering products or unions between certain known spaces, as spheres and projective spaces. Moreover, in Toda's work, it is shown that it is possible to construct a space of type (a, b) for any choice of a and b. For example,

1. The product $S^n \times S^{2n}$ is a space of the type $(0, 1)$, characterized by n.

2. The one point union $S^n \vee S^{2n} \vee S^{3n}$ is of type $(0, 0)$, characterized by n.

3. The one point union $S^6 \vee \mathbb{C}P^2$ is of type $(1, 0)$, characterized by $n = 2$.

4. The projective spaces $\mathbb{R}P^3, \mathbb{C}P^3, \mathbb{H}P^3$ are examples of spaces of type $(1, 1)$, characterized by $n = 1, n = 2$ and $n = 3$, respectively.

In 2010, Pergher et al. [20] investigated the existence of free actions of the groups $\mathbb{Z}_2$ and S^1 on spaces of type (a, b), for $n > 1$, where they concluded that:

Theorem 3.2.1. [20] Let X be a space of type (a, b), characterized by $n > 1$.

i. If a is odd and b is even, then $\mathbb{Z}_2$ cannot act freely on X.

ii. If $a \neq 0$, then S^1 cannot act freely on X.

iii. If $G = \mathbb{Z}_2$ act freely on X where both a and b even, then $H^*(X/G;\mathbb{Z}_2) \cong \mathbb{Z}_2[x, z]/\langle x^{3n+1}, z^2, zx^{n+1} \rangle$, where $\deg x = 1$ and $\deg z = n$.

iv. If $G = S^1$ act freely on X, then $a = 0$ and $H^*(X/G;\mathbb{Z}_2)$ is isomorphic to one of the following graded commutative algebras

$$\mathbb{Z}_2[x, z]/\left\langle x^{(3n+1)/2}, z^2, zx^{(n+1)/2} \right\rangle, \text{ where } \deg x = 2 \text{ and } \deg z = n,$$

or

$$\mathbb{Z}_2[x, z]/\left\langle x^{(n+1)/2}, z^2 \right\rangle, \text{where } \deg x = 2, \deg z = 2n \text{ and } b \text{ is odd.}$$

In [21], Dotzel and Singh constructed a class of examples of free actions of $\mathbb{Z}_p, p$ prime, on spaces of type $(0, 0)$, by using some known topological operations.

In particular, for n even, it is shown in [22] that the only group that can act freely on X is $\mathbb{Z}_2$. In addiction, the authors construct a example of such action. If n is odd and X is of type $(0, 1)$, then any finite group G which acts freely on X cannot contain the group $\mathbb{Z}_p \oplus \mathbb{Z}_p$, for any p odd prime.

Theorem 3.2.2. Let X be a manifold[5] that is a space of type $(0, b)$, characterized by $n > 1$. If $G = S^3$ acts freely on X, then n is an odd number of the form $4k - 1$, for some $k \geq 1$ and b is odd. In this case, the cohomology algebra of the orbit space X/G is isomorphic to the graded polynomial algebra $\mathbb{Z}_2[x, y]/\langle x^k, y^2 \rangle$, where $\deg x = 4$ and $\deg y = 2n$.

Proof. Let us suppose know that the group $G = S^3$ acts freely on a space X of type $(0, b)$ and let $\{E_r^{*,*}, d_r\}$ be the spectral sequence associated with the Borel fibration $X_G \to B_G$, with fiber X, such that $E_2^{p,q} = H^p(B_G; \mathbb{Z}_2) \otimes H^q(X; \mathbb{Z}_2)$, which converges to $H^*(X_G; \mathbb{Z}_2) \cong H^*(X/G; \mathbb{Z}_2)$.

By Quotient Manifold Theorem, it follows that this sequence does not collapse on its E_2−term. Then, there must exist some nontrivial differential d_{r_i}, for some $r_i \geq 2$. If $r = \min\{r_i\}$, then

$$E_r^{p,q} \cong E_{r-1}^{p,q} \cong \cdots \cong E_2^{p,q},$$

and this is possible only if $r = 4k$ and $n = 4k - 1$, for some $k \geq 1$. This provides the following possibilities for the action of the differentials $d_{4k}^{4l,q}$, for $q = n, 2n, 3n$:

a. $d_r(1 \otimes u_1) = 0$, $d_r(1 \otimes u_2) = \tau^k \otimes u_1$ and $d_r(1 \otimes u_3) = \tau^k \otimes u_2$,

b. $d_r(1 \otimes u_1) = \tau^k \otimes 1$ $d_r(1 \otimes u_2) = \tau^k \otimes u_1$ and $d_r(1 \otimes u_3) = 0$,

c. $d_r(1 \otimes u_1) = \tau^k \otimes 1$, $d_r(1 \otimes u_2) = 0$ and $d_r(1 \otimes u_3) = \tau^k \otimes u_2$,

d. $d_r(1 \otimes u_1) = 0$, $d_r(1 \otimes u_2) = \tau^k \otimes u_1$ and $d_r(1 \otimes u_3) = \tau^k \otimes u_2$,

e. $d_r(1 \otimes u_1) = 0$, $d_r(1 \otimes u_2) = 0$ and $d_r(1 \otimes u_3) = \tau^k \otimes u_2$,

f. $d_r(1 \otimes u_1) = 0$, $d_r(1 \otimes u_2) = \tau^k \otimes u_1$ and $d_r(1 \otimes u_3) = 0$,

g. $d_r(1 \otimes u_1) = \tau^k \otimes 1$, $d_r(1 \otimes u_2) = 0$ and $d_r(1 \otimes u_3) = 0$.

We will divide the analysis of these cases according to the parity of b.

Case b odd: In this case, we have the relation $u_1 u_2 = u_3$ and, by the multiplicative properties of the differentials, we have

---

[5] This proof works even if X is a more general finitistic space of type (a, b), and for that we only need an adaptation of the Quotient Manifold Theorem for a more general result concerning the cohomology of the quotient of a finitistic CW complex space. More precisely, it is possible to show that if X is a finitistic free G−space, where $G = \mathbb{Z}_2, S^1$ or $G = S^3$, and if there is $n > 0$ such that $H^j(X; \mathbb{Z}_2) = \{0\}$ for all $j > n$ then $H^j(X/G; \mathbb{Z}_2) = \{0\}$ for all $j > n$.

$$d_r(1\otimes u_3) = (1\otimes u_1)d_r(1\otimes u_2) + (1\otimes u_2)d_r(1\otimes u_1).$$

So, if one of the cases (b),(d),(e), or (g) occurred, it would lead to the contradiction $00 = \tau^k \otimes u_2$.

If case (a) occurred, then the differentials $d_{4k}^{4i,3n}$ and $d_{4k}^{4i,2n}$ would be isomorphisms, whence it would follow that

$$\text{im} d_{4k}^{4i,3n} \cong \mathbb{Z}_2 \not\subseteq \{0\} = \text{ker} d_{4k}^{4(i+k),2n},$$

which is a contradiction.

If case (f) occurred, then the sequence would collapse on its E_{4k+1}–term, with the lines $E_{4k+1}^{*,0}$ and $E_{4k+1}^{*,3n}$ containing an infinite number of nonzero elements. This would contradict the Quotient Manifold Theorem.

Therefore, (c) is the only possible case, and it produces the following pattern:

$$E_{4k+1}^{p,q} = \begin{cases} \mathbb{Z}_2, & \text{if } p = 0, 4, \cdots, 4(k-1) \text{ and } q = 2n, \\ \{0\}, & \text{otherwise.} \end{cases}$$

Then, the sequence collapses on its E_{4k+1}–term, and $E_\infty^{p,q} \cong E_{4k+1}^{p,q}$, for all $p,q \geq 0$. So, $H^j(X/G;\mathbb{Z}_2) \cong \text{Tot}^j(E_\infty)$.

The elements $\tau\otimes 1$ and $1\otimes u_2$ are the only permanent co-cycles, so they determine the nonzero elements x and y in $E_\infty^{4,0}$ and $E_\infty^{0,2n}$, respectively. By (15), we have $\pi^*(\tau) = x$; then, $0 = \pi^*(\tau^j) = x^j$ for all $j \geq k$. By the structure of the E_∞–term, it follows that $y^2 = 0$; therefore, $H^*(X/G;\mathbb{Z}_2)$ is isomorphic to the graded polynomial algebra $\mathbb{Z}_2[x,y]/\langle x^k, y^2\rangle$, where $\deg x = 4$ and $\deg y = 2n$.

Case b even: We will show that if b is even, then none of the cases can occur. By the relation $u_1u_2 = 0$, we have

$$0 = (1\otimes u_1)d_r(1\otimes u_2) + (1\otimes u_2)d_r(1\otimes u_1),$$

and this allows us to eliminate the cases (b),(c), and (g), since they produce the contradiction $0 = \tau^k \otimes u_2$.

By the same reason of the previous case (b odd), we can eliminate case (a); that is, it implies that

$$\text{im} d_{4k}^{4i,3n} \not\subseteq \text{ker} d_{4k}^{4(i+k),2n}.$$

By a similar reason we can eliminate (d), since it implies that the differentials $d_{4k}^{4i,2n}$ and $d_{4k}^{4j,3n}$ are isomorphisms.

For case (e), the sequence would collapse on its E_{4k+1}–term, with the lines $E_{4k}^{*,0}$ and $E_{4k}^{*,n}$ containing infinite nonzero elements, which would contradict the Quotient Manifold Theorem. Finally, by the same reason of the previous case, we can eliminate (f); therefore, when b is even, the space X does not admit any free action of G. □

3.3 Free actions on Dold, Wall, and Milnor manifolds

The Dold manifolds $P(m,n)$, as they came to be known, were defined by A. Dold [23] as orbit spaces of free actions of $\mathbb{Z}_2$, or equivalently free involutions, on a product of the form $S^m \times \mathbb{C}P^n$. Precisely, for each pair of nonnegative integers m and n, $P(m, n)$ is the orbit space $S^m \times \mathbb{C}P^n/T$, where $T(x, [z]) = (-x, [\bar{z}])$.

Let $R : S^m \to S^m$ be the involution defined by the reflection of the last coordinate $R(x_0, \cdots, x_m) = (x_0, \cdots, x_{m-1}, -x_m)$, and $1 : \mathbb{C}P^n \to \mathbb{C}P^n$ be the identity map. Since the involution $R \times 1 : S^m \times \mathbb{C}P^n \to S^m \times \mathbb{C}P^n$ commutes with the involution T, it induces an involution $S : P(m, n) \to P(m, n)$.

For each pair of nonnegative integers m and n, the Wall manifold[6] $Q(m, n)$ is defined as the mapping torus of the homeomorphism S, that is,

$$Q(m, n) = \frac{P(m, n) \times [0, 1]}{([x, z], 0) \sim (S[x, z], 1)}. \tag{23}$$

Let m, n be integers, such that $0 \leq n \leq m$. It is called a (real) Milnor manifold[7] of dimensions $n + m - 1$ to the smooth closed submanifold of codimension 1 in $\mathbb{R}P^m \times \mathbb{R}P^n$, described in homogeneous coordinates as:

$$\mathbb{R}H_{m,n} = \{([x_0, \cdots, x_m], [y_0, \cdots, y_n]) \in \mathbb{R}P^m \times \mathbb{R}P^n \,|\, x_0 y_0 + \cdots + x_n y_n = 0\}, \tag{24}$$

which is also denoted by $H(m, n)$. Equivalently, $\mathbb{R}H_{m,n}$ is the total space of the bundle:

$$\mathbb{R}P^{m-1} \hookrightarrow \mathbb{R}H_{m,n} \xrightarrow{\pi} \mathbb{R}P^n. \tag{25}$$

The manifolds $P(m, n)$, $Q(m, n)$, and $H(m, n)$ were constructed to provide representatives for generators in odd dimension to the unoriented cobordism ring $\mathfrak{R}_*$, since we have the projective spaces as representatives in even dimensions. Precisely, the following sets are generator sets for $\mathfrak{R}_*$:

$$\{[\mathbb{R}P^{2i}], [P(2^r - 1, s2^r)]; i, r, s \geq 1\}, \tag{26}$$

$$\{[\mathbb{R}P^{2i}], [Q(2^r - 2, s2^r)]; i, r, s \geq 1\}, \tag{27}$$

and

$$\{[\mathbb{R}P^{2i}], [H(2^k, 2t2^k)]; i, k, t \geq 1\}. \tag{28}$$

For these reasons, the analysis of certain structures and algebraic invariants related to the Dold, Milnor, and Wall manifolds is a relevant research topic, as it is done on the works [26–28] of Mukerjee. On the particular interest of investigating the existence of free actions of compact Lie groups on these spaces and also the cohomology classification of the respective orbit spaces, there are several results in the literature in which we will briefly discuss some of them below.

Regarding the existence of free actions of $\mathbb{Z}_2$ on Dold manifolds, Morita et al. [29] partially solved the problem by considering free involutions on $P(1, n)$, for $n \geq 1$ an odd integer. Later this problem was completely solved by Dey [30], according to the following.

Theorem 3.3.1. [30] If $G = \mathbb{Z}_2$ acts freely on $X = P(m, n)$, then $H^*(X/G; \mathbb{Z}_2)$ is isomorphic to one of the following graded algebras:

6 For more details on the construction of this manifolds, see [24].

7 For more details on Milnor's manifold, see [25, 26].

i. (*i*) $\mathbb{Z}_2[x, y, z]/\langle x^2, y^{(m+1)/2}, z^{n+1}\rangle$, where $\deg x = 1$, $\deg y = \deg z = 2$, and m is odd.

ii. $\mathbb{Z}_2[x, y, z]/\langle f, g, z^{(n+1)/2} + h\rangle$, where n is odd, $\deg x = \deg y = 1$, $\deg z = 4$, $f = (x^{m+1} + \alpha_1 x^m y + \alpha_2 x^{m-1} y^2)$, and $g = (y^3 + \beta_1 x y^2 + \beta_2 x^2 y)$, with $\alpha_i, \beta_i \in \mathbb{Z}_2$, and $h \in \mathbb{Z}_2[x, y, z]$ is either the zero polynomial or it is a homogeneous polynomial of degree $2n + 2$ with the highest power of z less than or equal to $(n-1)/2$.

Concerning on free involutions on Wall manifods $Q(m, n)$, the work of Khare [31] shows that these manifolds bounds if and only if n is odd or $n = 0$ and m odd. By Proposition 3.5 in [32], we can conclude that $X = Q(m, \text{odd})$ admit free involutions and about the orbit spaces $X/\mathbb{Z}_2$ we have the following result, for some values of m.

Theorem 3.3.2. [32, 33] Let $X = Q(m, n)$, where $n > 0$ is odd, equipped with a free action of the group $G = \mathbb{Z}_2$.

i. If $m = 1$ and the induced action of $\mathbb{Z}_2$ on the mod 2 cohomology is trivial, then

$$H^*(X/G; \mathbb{Z}_2) \cong \mathbb{Z}_2[x, y, z, w]/\left\langle x^3, y^3, z^2, y^2 + yw, w^{(n+1)/2}\right\rangle,$$

where $\deg x = \deg y = \deg z = 1$, and $\deg w = 4$.

ii. If m is even and the induced action of $\mathbb{Z}_2$ on the mod 2 cohomology is trivial, then $H^*(X/\mathbb{Z}_2)$ is isomorphic to one of the following graded polynomial algebras:

$$\mathbb{Z}_2[x, y, z, w]/\left\langle x^3, y^2, z^{m+1} + z^m y, w^{(n+1)/2}\right\rangle,$$

where $\deg x = \deg y = \deg z = 1$ and $\deg w = 4$, or

$$\mathbb{Z}_2[x, y, z, w]/\langle x^2, y^{m+1}, y^2 + z, w^{n+1}\rangle,$$

where $\deg x = \deg y = 1$ and $\deg z = \deg w = 2$.

Example 3.3.1. (Free S^1−actions on Dold manifolds) Let G be the group S^1 and m,n odd integers, where $m = 2k - 1$, for some $k \geq 1$. Considering $S^m \subseteq \mathbb{C}^k$, we define a free action of G on $S^m \times \mathbb{C}P^n$ by:

$$z * (w, v) \mapsto ((zw_1, \cdots, zw_k), [zv_0 : \cdots : zv_n]), \tag{29}$$

where $w = (w_1, \cdots, w_k) \in S^m \subseteq \mathbb{C}^k$ and $v = [v_0 : \cdots : v_n] \in \mathbb{C}P^n$.

Let us consider an arbitrary element $[w, v] \in P(m, n) = (S^m \times \mathbb{C}P^n)/T$, and note that the isotropy subgroup $G_{[w,\, v]}$ is trivial. Therefore, z must be equal to $1 \in G$, that is, $G_{[w,\, v]} = \{1\}$, so the induced action on Dold manifold $P(m, n)$ is free.

If n is odd and $m = 2k$ is even, then we can consider

$$S^m = \left\{(w,\ t) \in \mathbb{C}^k \times \mathbb{R}; \|w\| + |t| = 1\right\}, \tag{30}$$

and the analogous free action of G on $S^m \times \mathbb{C}P^n$ is defined by:

$$z * ((w,\ t), v) \mapsto ((zw_1,\ \cdots,\ zw_k,\ t),\ [zv_0 : \cdots : zv_n]). \tag{31}$$

Since this action is free, it induces a free action of G on $P(m, n)$, as in the previous case.

Theorem 3.3.3. There is no free action of $G = S^1$ on $X = Q(m, n)$, for any $m,n > 0$.

Proof. Recall that[8] $H^*(X;\mathbb{Z}_2) \cong \mathbb{Z}_2[x, c, d]/\langle x^2, c^{m+1} + c^m x, d^{n+1}\rangle$, where $\deg x = \deg c = 1$ and $\deg d = 2$. Suppose that there is a free action of $G = S^1$ on X and let $X \hookrightarrow X_G \to B_G$ be the associated Borel fibration, with $\{E_r^{*,*}, d_r\}$, and such that $E_2^{p,q} \cong H^p(B_G;\mathbb{Z}_2) \otimes H^q(X;\mathbb{Z}_2)$, that converges, as an algebra, to $H^*(X_G;\mathbb{Z}_2) \cong H^*(X/G;\mathbb{Z}_2)$. Since this sequence does not collapse on E_2−page, there is some nontrivial differential $d_2^{*,*}$, according to the following cases.

Case m odd: In this case, we have $d_2^{0,m+1}(1 \otimes c^{m+1}) = 0$. In fact, since $m + 1 = 2r$, for some $r > 0$, then $d_2^{0,2r}(1 \otimes c^{2r}) = d_2^{0,2r}((1 \otimes c^r)(1 \otimes c^r)) = 0$. However, by relation $c^{m+1} = c^m x$, it follows that

$$d_2^{0,m+1}(1 \otimes c^{m+1}) = (1 \otimes c^m) d_2^{0,1}(1 \otimes x) + (1 \otimes x) d_2^{0,m}(1 \otimes c^m).$$

Therefore, $d_2^{0,1}(1 \otimes x) \neq 0$ and $d_2^{0,1}(1 \otimes c) \neq 0$, cannot occur simultaneously because in this case we will have $0 = \tau \otimes c^{m-1}(c + x)$, which is a contradiction.

Similarly, $d_2^{0,1}(1 \otimes x) = 0$ and $d_2^{0,1}(1 \otimes c) \neq 0$, cannot occur simultaneously. Therefore, it follows there are only the possibilities:

1.$d_2^{0,1}(1 \otimes c) = d_2^{0,1}(1 \otimes x) = 0$ and $d_2^{0,2}(1 \otimes d) = \tau \otimes c$,

2.$d_2^{0,1}(1 \otimes c) = d_2^{0,1}(1 \otimes x) = 0$ and $d_2^{0,2}(1 \otimes d) = \tau \otimes x$.

We claim that (1) and (2) cannot occur.

If (1) occurs, then for any $j \geq 0$, $k \in \{1\cdots, m\}$ and $l \in \{1\cdots, n\}$, it follows that

$$d_2^{j,2l}\left(\tau^j \otimes d^l\right) = \begin{cases} 0, & \text{if } l \text{ is even}, \\ \tau^{j+1} \otimes c d^{l-1}, & \text{if } l \text{ is odd}, \end{cases}$$

$$d_2^{j,k+2l}\left(\tau^j \otimes c^k d^l\right) = \begin{cases} 0, & \text{if } l \text{ is even}, \\ \tau^{j+1} \otimes c^{k+1} d^{l-1}, & \text{if } l \text{ is odd}, \end{cases}$$

$$d_2^{j,2l+1}\left(\tau^j \otimes x d^l\right) = \begin{cases} 0, & \text{if } l \text{ is even}, \\ \tau^{j+1} \otimes c x d^{l-1}, & \text{if } l \text{ is odd}, \end{cases}$$

$$d_2^{j,2l+k+1}\left(\tau^j \otimes x c^k d^l\right) = \begin{cases} 0, & \text{if } l \text{ is even}, \\ \tau^{j+1} \otimes x c^{k+1} d^{l-1}, & \text{if } l \text{ is odd}, \end{cases}$$

therefore, we will have $E_3^{p,q} \cong \{0\}$, for all p odd or $q \equiv 2(\text{mod}4)$ and $q > 0$. We can see that the sequence collapses on E_3−page, however $E_3^{2r,q} \neq \{0\}$, for all $q \equiv s(\text{mod}4)$, $s = 0,2,3$, and $r \geq 0$, which contradicts the Quotient Manifold Theorem.

[8] See [24].

If the case (2) occurs, then for all $j \geq 0$, $k \in \{1\cdots, m\}$ and $l \in \{1\cdots, n\}$, we have

$$d_2^{j,2l}\left(\tau^j \otimes d^l\right) = \begin{cases} 0, & \text{if } l \text{ is even,} \\ \tau^{j+1} \otimes xd^{l-1}, & \text{if } l \text{ is odd,} \end{cases}$$

$$d_2^{j,2l+k}\left(\tau^j \otimes c^k d^l\right) = \begin{cases} 0, & \text{if } l \text{ is even,} \\ \tau^{j+1} \otimes xc^k d^{l-1}, & \text{if } l \text{ is odd,} \end{cases}$$

while $d_2^{j,2l+1}\left(\tau^j \otimes xd^l\right) = 0$, since $x^2 = 0$. Therefore, analogous to case (1), we can conclude that this case is not possible either.

Case m even: In this case, we have $d_2^{0,m}(1 \otimes c^m) = 0$, so $d_2^{0,m+1}(1 \otimes c^m x) = (1 \otimes c^m) d_2^{0,1}(1 \otimes x)$, while, by relation $c^{m+1} = c^m x$, $d_2^{0,m+1}(1 \otimes c^m x) = d_2^{0,m+1}(1 \otimes c^{m+1}) = (1 \otimes c^m) d_2^{0,1}(1 \otimes c)$, therefore, we should have necessarily $d_2^{0,1}(1 \otimes c) = d_2^{0,1}(1 \otimes x)$.

If $d_2^{0,1}(1 \otimes c) = d_2^{0,1}(1 \otimes x) = \tau \otimes 1$, then $d_2^{0,2}(1 \otimes d) = 0$, otherwise we will have

$$\operatorname{im} d_2^{0,2} = \langle d_2^{0,2}(1 \otimes d) \rangle \not\subset \ker d_2^{2,1} \langle \tau \otimes (c + x) \rangle,$$

which is a contradiction.

Let us suppose now that $d_2^{0,1}(1 \otimes c)$ and $d_2^{0,1}(1 \otimes x)$ are nontrivial, and $d_2^{0,2}(1 \otimes d) = \tau \otimes c$. Then, for example,

$$\operatorname{im} d_2^{0,4} = \langle \tau \otimes cd \rangle \oplus \langle \tau \otimes c^2 x \rangle \not\subset \ker d_2^{2,3} = \langle \tau \otimes xc^2 \rangle \oplus \langle \tau \otimes c^3 \rangle,$$

which is a contradiction. Therefore, for m even, we must consider only the cases:

a. $d_2^{0,1}(1 \otimes c) = d_2^{0,1}(1 \otimes x) = \tau \otimes 1$ and $d_2^{0,2}(1 \otimes d) = 0$;

i. $d_2^{0,1}(1 \otimes c) = d_2^{0,1}(1 \otimes x) = 0$ and $d_2^{0,2}(1 \otimes d) = \tau \otimes x$.

We will show that both (i) and (ii) cannot occur.

If (i) is true, then for all $j \geq 0$, $k \in \{1\cdots, m\}$ and all $l \in \{1\cdots, n\}$, we have

$$d_2^{j,k}\left(\tau^j \otimes c^k\right) = \begin{cases} 0, & \text{if } k \text{ is even,} \\ \tau^{j+1} \otimes c^{k-1}, & \text{if } k \text{ is odd,} \end{cases}$$

$$d_2^{j,k+1}\left(\tau^j \otimes c^k x\right) = \begin{cases} \tau^{j+1} \otimes c^k, & \text{if } k \text{ is even,} \\ \tau^{j+1} \otimes (c + x) c^{k-1}, & \text{if } k \text{ is odd,} \end{cases}$$

$$d_2^{j,2l+k}\left(\tau^j \otimes c^k d^l\right) = \begin{cases} 0, & \text{if } k \text{ is even,} \\ \tau^{j+1} \otimes c^{k-1} d^l, & \text{if } k \text{ is odd,} \end{cases}$$

and $d_2^{j,2l+1}\left(\tau^j \otimes xd^l\right) = \tau^{j+1} \otimes d^l$. Therefore, $E_3^{p,q} \cong \{0\}$, for all $q \not\equiv 1(\text{mod} 4)$, e $p > 0$. However, for $q \equiv 1(\text{mod} 4)$, $q \geq 5$, we have

$$E_3^{2j,q} \cong \left\langle \tau^j \otimes (c + x) d^{(q-1)/2} \right\rangle \neq \{0\}.$$

which contradicts the Quotient Manifold Theorem.

For (iii), note that it will result in a pattern similar to case (2); therefore, it cannot occur either due to the same arguments. □

In order to investigate the existence of free involutions on a Milnor manifold $X = H(m, n)$, Dey and Singh [34] showed that if $G = \mathbb{Z}_2$ acts freely on X, with $1 < n < m$ and $m \not\equiv 2(\mathrm{mod}4)$, then necessarily m and n must be odd. Furthermore, they construct some examples of such free actions and, in this case, it follows that

$$H^*(X/G; \mathbb{Z}_2) \cong \mathbb{Z}_2[x, y, z, w]/I, \tag{32}$$

where

$$I = \langle z^2, w^2 - \gamma_1 zw - \gamma_2 x - \gamma_3 y, x^{(n+1)/2} + \alpha_0 zwx^{(n-1)/2}y + \cdots + \alpha_{\frac{n-1}{2}} zwy^{(n-1)/2},$$
$$(w + \beta_0 z)y^{(m-1)/2} + (w + \beta_1 z)xy^{(m-3)/2} + \cdots + \left(w + \beta_{\frac{n-1}{2}} z\right)x^{(n-1)/2}y^{(m-n)/2}\rangle,$$

with $\deg x = \deg y = 2$, $\deg z = \deg w = 1$, and $\alpha_i, \beta_i, \gamma_i \in \mathbb{Z}_2$.

If $G = S^1$ act freely on $X = H(m, n)$, then $H^*(X/G; \mathbb{Z}_2) \cong \mathbb{Z}_2[x, y, w]/I$, where

$$I = \left\langle x^{(n+1)/2}, wy^{(m-1)/2} + xwy^{(m-3)/2} + \cdots + wx^{(n-1)/2}y^{(m-n)/2}, w^2 - \alpha x - \beta y \right\rangle,$$

with $\deg x = \deg y = 2$, $\deg w = 1$, and $\alpha, \beta \in \mathbb{Z}_2$.

Author details

Thales Fernando Vilamaior Paiva
Universidade Federal de Mato Grosso do Sul, Aquidauana, Brazil

*Address all correspondence to: thales.paiva@ufms.br

References

[1] Zum HH. Clifford-Kleinschen Raumproblem. Mathematische Annalen. 1926;**95**(1):313-339

[2] Bredon G. Introduction to Compact Transformation Groups. New York: Academic Press; 1972

[3] de Medrano SL. Involutions on Manifolds. In: Ergebnisse der Mathematik und ihrer Grenzgebiete 59. Springer; 1971

[4] Lee JM. Introduction to Smooth Manifolds. 2nd ed. New York: Springer-Verlag; 2012

[5] Milnor J. Construction of Universal Bundles. II. Annals of Mathematics. 1956: 430-436

[6] Husemoller D. Fibre Bundles. 3rd ed. New York: Springer-Verlag; 1994

[7] Whitehead GW. Elements of Homotopy Theory. New York: Springer-Verlag; 1978

[8] Dieck T. Transformation Groups. Berlin: Walter de Gruyter; 1987

[9] Mccleary J. A User's Guide to Spectral Sequences. 2nd ed. New York: Cambridge University Press; 2001

[10] Deo S, Tripathi HS. Compact Lie group actions on finitistic spaces. Topology. 1982;**21**:393-399

[11] Deo S, Singh TB. On the converse of some theorems about orbit spaces. Journal of the London Mathematical Society. 1982;**(1) 25**:162-170

[12] Cusick LW. Finite groups that can act freely on products of even dimensional spheres. Indiana University Mathematics Journal. 1986;**35**(1):175-178

[13] Dotzel RM, Singh TB, Tripathi SP. The cohomology rings of the orbit spaces of free transformation groups of the product of two spheres. Proceedings of American Mathematical Society. 2001; **129**(3):921-930

[14] Kumari A, Singh HK. Fixed point free actions of spheres and equivariant maps. Topology and its Applications. 2022;**305**:107886

[15] Singh M. Cohomology algebra of orbit spaces of free involutions on lens spaces. Journal of the Mathematical Society of Japan. 2013;**65**(4):1055-1078

[16] Singh M. Orbit spaces of free involutions on the product of two projective spaces. Results in Mathematics. 2010;**57**:53-67

[17] Kaur J, Singh HK. On the existence of free actions of S^3 on certain finitistic mod p cohomology spaces. Journal of the Indian Mathematical Society. 2015;**82**: 97-106

[18] James I. Notes on cup products. Proceedings of the American Mathematical Society. 1957;**8**:374-383

[19] Toda H. Note on cohomology ring of certain spaces. Proceedings of the American Mathematical Society. 1963; **14**:89-95

[20] Pergher PLQ, Singh HK, Singh TB. On $\mathbb{Z}_2$ and S^1 free actions on spaces of cohomology type (a, b). Houston Journal of Mathematics. 2010;**36**(1):137-146

[21] Dotzel RM, Singh TB. Z_p actions on spaces of cohomology type $(a, 0)$. Proceedings of the American Mathematical Society. 1991;**113**(3): 875-878

[22] Singh KS, Singh HK, Singh TB. $(0, b)$. Glasgow Mathematical Journal. 2018;**60**:673-680

[23] Dold A. Erzeugende der Thomschen Algebra $\mathfrak{N}$. Mathematische Zeitschrift. 1956;**65**:25-35

[24] Wall CTC. Determination of the cobordism ring. Annals of Mathematics. 1960;**72**:292-311

[25] Milnor J. On the Stiefel-Whitney numbers of complex manifolds and spin manifolds. Topology. 1965;**3**:223-230

[26] Mukerjee HK. Classification of homotopy real Milnor manifolds. Topology and its Applications. 2004;**139**: 151-184

[27] Mukerjee HK. Classification of homotopy Dold manifolds. New York Journal of Mathematics. 2003;**9**:1-23

[28] Mukerjee HK. Classification of homotopy Wall's manifolds. Topogy and its Applications. 2006;**153**: 3467-3495

[29] Morita AMM, Mattos D, Pergher PLQ. The cohomology ring of orbit spaces of free $\mathbb{Z}_2-$actions on some Dold manifolds. Bulletin of the Australian Mathematical Society. 2018; **97**:340-348

[30] Dey P. Free actions of finite groups on products of Dold manifolds. arXiv: 1809.02307, 12-Apr-2019

[31] Khare SS. On bounding and independence of wall manifolds. Journal of the Indian Mathematical Society. 2004;**71**:195-201

[32] Paiva TFV, dos Santos EL. Cohomology algebra of orbit spaces of free involutions on some wall manifolds. Bulletin of the Brazilian Mathematical Society. Bulletin of the Brazilian Mathematical Society, New Series. 2022; **53**:281-294

[33] Paiva TFV, dos Santos EL. Free involution on odd dimensional Wall manifolds and cohomology of orbit spaces. Topology and Its Applications. 2022

[34] Dey P, Singh M. Free actions of some compact groups on Milnor manifolds. Glasgow Mathematical Journal. 2019;**61**: 727-742

Chapter 2

Perspective Chapter: Quasi Conformally Flat Quasi Einstein-Weyl Manifolds

Fusun Nurcan

Abstract

The aim of this work is to study on quasi conformally flat quasi Einstein-Weyl manifolds. In this book chapter, firstly, an interesting relationship between complementary vector field and generator of the quasi Einstein-Weyl manifold is obtained and supported by an example. Then, it is investigated that quasi conformally flat quasi Einstein-Weyl manifolds are of quasi constant curvature, recurrent and semi-symmetric under which conditions after obtaining the expression of the curvature tensor of the quasi conformally flat quasi Einstein-Weyl manifold. Furthermore, some equivalences are obtained between to be of quasi constant curvature and to be semi-symmetric in quasi conformally flat quasi Einstein-Weyl manifolds.

Keywords: quasi Einstein-Weyl manifold, Weyl manifold of quasi constant curvature, quasi conformally flat manifold, recurrent manifold, semi-symmetric manifold

1. Introduction

In 1918, H. Weyl generalized Riemannian geometry as a new way to formulate the unified field theory in physics and defined Weyl manifolds with conformal metric and symmetric connection [1]. After this study, Weyl manifolds attracted the attention of many mathematicians. In 1943, E. Cartan defined Einstein-Weyl manifolds and studied three-dimensional Einstein-Weyl spaces [2]. In 1985, P.E. Jones and K.P. Tod have studied Einstein-Weyl spaces, and then they have done many studies on this subject [3]. Although Weyl's theory did not attract much attention in physics, it attracted the attention of mathematicians and studies have been carried out on this subject until today.

An n-dimensional Weyl manifold M is defined as a manifold with a torsion-free connection Γ and a conformal metric tensor g_{ij}, if the compatible condition is in the form of

$$\nabla_k g_{ij} - 2g_{ij}\Phi_k = 0 \tag{1}$$

which is equivalent to

$$\nabla_k g^{ij} + 2g^{ij}\Phi_k = 0, \tag{2}$$

where Φ_k is a complementary covariant vector field [4]. Such a Weyl manifold is denoted by $\left(M, g_{ij}, \Phi_k\right)$ and (1) tells us that a Riemannian manifold is obtained if $\Phi_k = 0$ or Φ_k is gradient.

Φ_k changes by

$$\tilde{\Phi}_k = \Phi_k + \partial_k(\log \lambda) \tag{3}$$

under the transformation of the metric tensor g_{ij} in the form of

$$\tilde{g}_{ij} = \lambda^2 g_{ij} \tag{4}$$

where λ is a point function [4]. With reference to this transformation, the quantity A is called a satellite of g_{ij} with the weight of $\{p\}$ if it changes by [5]

$$\tilde{A} = \lambda^p A \tag{5}$$

and the quantity $\dot{\nabla}_k A$ is called prolonged covariant derivative of the satellite A of g_{ij} with the weight of $\{p\}$ if it is defined by [5]

$$\dot{\nabla}_k A = \nabla_k A - p\Phi_k A. \tag{6}$$

From (1), (4) and (6), we have

$$\dot{\nabla}_k g_{ij} = 0 \tag{7}$$

which g_{ij} is with the weight of $\{2\}$.

The coefficients Γ^i_{jk}'s of a torsion-free connection Γ on the Weyl manifold $\left(M, g_{ij}, \Phi_k\right)$ are given by

$$\Gamma^i_{jk} = \left\{ \begin{matrix} i \\ jk \end{matrix} \right\} - \left(\delta^i_j \Phi_k + \delta^i_k \Phi_j - g_{jk} g^{ih} \Phi_h\right) \tag{8}$$

where $\left\{ \begin{matrix} i \\ jk \end{matrix} \right\}$'s are the Christoffel symbols of second kind [4].

The curvature tensor R^h_{ijk} of the symmetric connection Γ on the Weyl manifold is defined by

$$R^h_{ijk} = \partial_j \Gamma^h_{ik} - \partial_k \Gamma^h_{ij} + \Gamma^h_{rj} \Gamma^r_{ik} - \Gamma^h_{rk} \Gamma^r_{ij}. \tag{9}$$

The Ricci tensor R_{ij}, which is defined by $R_{ij} = R^h_{ijh}$, satisfies

$$R_{[ij]} = n \nabla_{[i} \Phi_{j]} = \frac{1}{2} R^r_{rji}. \tag{10}$$

With the help of (9), the conformal curvature tensor C^h_{ijk} and the concircular curvature tensor $\tilde{C}^h_{ijk}$ of a torsion-free connection Γ on the Weyl manifold are expressed by

$$C^h_{ijk} = R^h_{ijk} - \frac{1}{n}\delta^h_i R^r_{rjk} + \frac{1}{n-2}\left(\delta^h_j R_{ik} - \delta^h_k R_{ij} + g_{ik}g^{hm}R_{mj} - g_{ij}g^{hm}R_{mk}\right)$$
$$- \frac{1}{n(n-2)}\left(\delta^h_j R^r_{rki} - \delta^h_k R^r_{rji} + g_{ik}g^{hm}R^r_{rjm} - g_{ij}g^{hm}R^r_{rkm}\right)$$
$$+ \frac{R}{(n-1)(n-2)}\left(\delta^h_k g_{ij} - \delta^h_j g_{ik}\right), \tag{11}$$

$$\tilde{C}^h_{ijk} = R^h_{ijk} - \frac{R}{n(n-1)}\left(\delta^h_k g_{ij} - \delta^h_j g_{ik}\right), \tag{12}$$

where R^h_{ijk}, R_{ij} and R denote the curvature tensor, the Ricci tensor and the scalar curvature of Γ, respectively [6, 7].

In 1968, Yano and Sawaki defined and studied a new curvature tensor called quasi conformal curvature tensor on a Riemannian manifold [8]. Similarly, the notion of quasi conformal curvature tensor W^h_{ijk} of type $(1, 3)$ on a Weyl manifold of dimension n $(n > 3)$ is introduced by [9]

$$W^h_{ijk} = -(n-2)bC^h_{ijk} + [a + (n-2)b]\tilde{C}^h_{ijk}, \tag{13}$$

where a, b are arbitrary constants not simultaneously zero, C^h_{ijk} and $\tilde{C}^h_{ijk}$ are conformal curvature tensor and concircular curvature tensor of type $(1, 3)$, respectively.

By substituting (11) and (12) in (13) the quasi conformal curvature tensor can be expressed by

$$W^h_{ijk} = aR^h_{ijk} + b\left\{\delta^h_k R_{ij} - \delta^h_j R_{ik} + g_{ij}g^{hm}R_{mk} - g_{ik}g^{hm}R_{mj}\right\}$$
$$+\frac{b}{n}\left\{(n-2)\delta^h_i R^r_{rjk} + \delta^h_j R^r_{rki} - \delta^h_k R^r_{rji} + g_{ik}g^{hm}R^r_{rjm} - g_{ij}g^{hm}R^r_{rkm}\right\}$$
$$-\frac{R}{n}\left\{\frac{a}{n-1} + 2b\right\}\left(\delta^h_k g_{ij} - \delta^h_j g_{ik}\right). \tag{14}$$

2. The concept of quasi conformally flatness on quasi Einstein-Weyl manifolds

Quasi Einstein manifolds occupy a large place in the mathematical literature. For instance, research on quasi-Einstein manifolds helps us to understand the global character of topological spaces. Beside mathematics, studies on quasi-Einstein manifolds gain meaning with applications to general relativity.

The concept of quasi Einstein manifold was firstly introduced by M. C. Chaki and R. K. Maity as follows [10]:

A non-flat Riemannian manifold $\left(M_n, g_{ij}\right)$ $(n > 2)$ is defined to be a quasi Einstein manifold if its Ricci tensor R_{ij} of type $(0, 2)$ is not identically zero and satisfies the condition

$$R_{ij} = \alpha g_{ij} + \beta A_i A_j, \tag{15}$$

where α, β are scalars of which $\beta \neq 0$ and A_i is a non-zero unit covariant vector field. In such an n-dimensional manifold which is denoted by $(QE)_n$; α, β are called associated scalars and A_i is called the generator of the manifold.

After Chaki and Maity, quasi Einstein manifolds are studied by many other authors. Moreover, in the articles [11–13] that inspired this study, conformal flatness and quasi conformal flatness were examined on quasi Einstein manifolds.

In this study, the concept of quasi conformal flatness on quasi Einstein manifolds were adapted to quasi Einstein-Weyl manifolds which was introduced by İ. Gül and E. Ö. Canfes as follows [14]:

Definition 1. *A non-flat Weyl manifold* $\left(M, g_{ij}, \Phi_k\right)$ *of dimension* $n(n>2)$ *is said to be a quasi Einstein-Weyl manifold if the symmetric part of its Ricci tensor* R_{ij} *of type* $(0,2)$ *is not identically zero and satisfies the condition*

$$R_{(ij)} = \alpha g_{ij} + \beta A_i A_j \tag{16}$$

where α *and* β *are scalars of weight* $\{-2\}$ *with* $\beta \neq 0$. *The scalars* α, β *are called "associated scalars" and the unit covariant vector* A_i *of weight* $\{1\}$ *is called "generator of the manifold". Such a manifold is denoted* $(QEW)_n$.

Therefore the aim of the present book chapter is to examine quasi conformally flat quasi Einstein-Weyl manifolds. It is organized as follows: In Section 1, the general information about Weyl manifolds are given. In Section 2, a theorem which shows the relationship between complementary vector field Φ_k and generator A_k of quasi Einstein-Weyl manifold $(QEW)_n$ is proved and the expression of the curvature tensor of the quasi conformally flat quasi Einstein-Weyl manifold is obtained. In Section 3, three basic concepts are defined on quasi conformally flat quasi Einstein-Weyl manifolds and the necessary and sufficient conditions for these concepts are emphasized.

By means of (10) and (16), Ricci tensor R_{ij} of $(QEW)_n$ is expressed by

$$R_{ij} = \alpha g_{ij} + \beta A_i A_j + n\nabla_{i}\Phi_{j]} \tag{17}$$

which implies

$$R = \alpha n + \beta. \tag{18}$$

From (17), we have

$$\begin{aligned} R_{ij,l} - \mu_l R_{ij} = {} & g_{ij}\{\alpha_{,l} - \alpha(\mu_l - 2\Phi_l)\} + \left(\beta_{,l} - \beta\mu_l\right)A_i A_j + \beta\left(A_{i,l}A_j + A_i A_{j,l}\right) + \\ & \frac{n}{2}\left[\left(\Phi_{j,il} - \Phi_{i,jl}\right) - \mu_l\left(\Phi_{j,i} - \Phi_{i,j}\right)\right]. \end{aligned} \tag{19}$$

Since

$$R_{,l} = \left(g^{ij}R_{ij}\right)_{,l} = -2\Phi_l g^{ij}R_{ij} + g^{ij}R_{ij_{,l}} \tag{20}$$

and A_i is normalized by the condition

$$g^{ij}A_i A_j = 1 \Leftrightarrow A^j A_j = 1 \tag{21}$$

it is found that by multiplying (19) by g^{ij}

$$R_{,l} - (\mu_l - 2\Phi_l)R = n\{\alpha_{,l} - \alpha(\mu_l - 2\Phi_l)\} + (\beta_{,l} - \beta\mu_l) + \beta(A_{i,l}A^i + A^j A_{j,l}). \tag{22}$$

By means of (18),

$$R_{,l} - \mu_l R = n(\alpha_{,l} - \mu_l\alpha) + (\beta_{,l} - \mu_l\beta). \tag{23}$$

We obtain that

$$\Phi_l(R - \alpha n) = \beta A_{i,l}A^i \tag{24}$$

where $\beta = R - \alpha n$.

Hence we have the following:

Theorem 1. *The complementary vector field Φ_i and the generator A_i of the quasi Einstein-Weyl manifold $(QEW)_n$ are related by*

$$A_{i,l}A^i = \Phi_l. \tag{25}$$

Although the first part of the following example was given to prove the existence of the quasi Einstein-Weyl manifold $(QEW)_n$ $(n > 2)$ in [14], the verification of Theorem 1 is made by the author of the present book chapter in the second part of the example.

Example 1. *A three dimensional Weyl manifold M_3 is equipped with a metric g_{ij} by*

$$ds^2 = g_{ij}dx^i dx^j = e^{x^1}\left[(dx^1)^2 + (dx^2)^2\right] + (dx^3)^2$$

and a 1-form Φ whose components Φ_k given by $\Phi = e^{x^1}dx^2 + dx^3$. The nonzero coefficients Γ^i_{jk} of a torsion-free connection Γ are [14]

$$\Gamma^1_{11} = \frac{1}{2}, \quad \Gamma^1_{12} = \Gamma^1_{21} = -e^{x^1}, \quad \Gamma^1_{13} = \Gamma^1_{31} = -1, \quad \Gamma^1_{22} = -\frac{1}{2},$$

$$\Gamma^2_{11} = e^{x^1}, \quad \Gamma^2_{12} = \Gamma^2_{21} = \frac{1}{2}, \quad \Gamma^2_{22} = -e^{x^1}, \quad \Gamma^2_{23} = \Gamma^2_{32} = -1, \quad \Gamma^2_{33} = 1,$$

$$\Gamma^3_{11} = \Gamma^3_{22} = e^{x^1}, \quad \Gamma^3_{23} = \Gamma^3_{32} = -e^{x^1}, \quad \Gamma^3_{33} = -1.$$

It is clear that $\left(M_3, g_{ij}, \Phi_k\right)$ is a Weyl manifold with the connection Γ satisfying the condition (1). An elementary calculation gives the following nonzero components of the Ricci tensor [14]:

$$R_{11} = e^{x^1}\left(1 + e^{x^1}\right), \quad R_{12} = -R_{21} = \frac{3}{2}e^{x^1}, \quad R_{22} = e^{x^1},$$

$$R_{23} = R_{32} = -e^{x^1}, \quad R_{33} = e^{x^1}.$$

Moreover, the components of the symmetric parts of the Ricci tensor R_{ij} and the scalar curvature R are [14]

$$R_{(11)} = e^{x^1}\left(1 + e^{x^1}\right), \quad R_{(22)} = e^{x^1}, \quad R_{(23)} = -e^{x^1}, \quad R_{(33)} = e^{x^1}, \quad R = 2\left(1 + e^{x^1}\right).$$

Therefore, by considering (15), we find that [14]

$$\alpha = 1 + e^{x^1} \ , \ \beta = -\left(1 + e^{x^1}\right) \ ,$$

$$A_1 = 0 \ , \ A_2 = \frac{e^{x^1}}{\sqrt{1 + e^{x^1}}} \ , \ A_3 = \frac{1}{\sqrt{1 + e^{x^1}}}$$

where $g^{ij}A_iA_j = 1$. *Thus,* $\left(M_3, g_{ij}, \Phi_k\right)$ *is a quasi-Einstein Weyl manifold.*
Now covariant derivatives of A_1, A_2 *and* A_3 *with respect to* $x^k (k = 1, 2, 3)$ *are as follows:*

$$A_{1,1} = 0 \ , \ A_{1,2} = 0 \ , \ A_{1,3} = 0;$$

$$A_{2,1} = \frac{e^{x^1}}{2(1 + e^{x^1})^{\frac{3}{2}}}, \ A_{2,2} = \frac{e^{2x^1} - e^{x^1}}{\sqrt{1 + e^{x^1}}} \ , \ A_{2,3} = 0; \tag{26}$$

$$A_{3,1} = \frac{-e^{x^1}}{2(1 + e^{x^1})^{\frac{3}{2}}}, \ A_{3,2} = \frac{2e^{x^1}}{\sqrt{1 + e^{x^1}}} \ , \ A_{3,3} = \frac{1 - e^{x^1}}{\sqrt{1 + e^{x^1}}}$$

On the other hand, the reciprocals of A_i*'s are*

$$A^1 = 0 \ , \ A^2 = A^3 = \frac{1}{\sqrt{1 + e^{x^1}}} \tag{27}$$

By substituting (26) and (27) in (25),

$$A_{1,1}A^1 + A_{2,1} \ A^2 + A_{3,1}A^3 = \Phi_1 \ ,$$

$$A_{1,2}A^1 + A_{2,2} \ A^2 + A_{3,2}A^3 = \Phi_2 \ ,$$

$$A_{1,3}A^1 + A_{2,3} \ A^2 + A_{3,3}A^3 = \Phi_3$$

are obtained.

A Weyl manifold $\left(M, g_{ij}, \Phi_k\right)$ $(n > 3)$ is called quasi conformally flat, if the quasi conformal curvature tensor W^h_{ijk} satisfy the condition

$$W^h_{ijk} = 0. \tag{28}$$

Now, let us suppose that $(QEW)_n$ $(n > 3)$ is quasi conformally flat with $a \neq 0$ and $b \neq 0$. Then from (14),

$$R^h_{ijk} = \frac{-b}{a}\left\{\delta^h_k R_{ij} - \delta^h_j R_{ik} + g_{ij}g^{hm}R_{mk} - g_{ik}g^{hm}R_{mj}\right\} \tag{29}$$

$$+\frac{-2b}{an}\left\{(n-2)\delta^h_i R_{[kj]} + \delta^h_j R_{[ik]} - \delta^h_k R_{[ij]} + g_{ik}g^{hm}R_{[mj]} - g_{ij}g^{hm}R_{[mk]}\right\}$$

$$+\frac{R}{an}\left\{\frac{a}{n-1} + 2b\right\}\left(\delta^h_k g_{ij} - \delta^h_j g_{ik}\right)$$

On the other hand, since it is assumed that the manifold is $(QEW)_n$, its Ricci tensor R_{ij} can be written as (17) which satisfies (18).

Substituting (10), (17) and (18) in (29), the curvature tensor R^h_{ijk} is obtained as

$$R^h_{ijk} = P\left(\delta^h_k g_{ij} - \delta^h_j g_{ik}\right) + Q\left\{\delta^h_k A_i A_j - \delta^h_j A_i A_k + g_{ij} g^{hm} A_m A_k - g_{ik} g^{hm} A_m A_j\right\} - \frac{(n-2)b}{a}\left\{2\delta^h_i \nabla_k \Phi_{j]} + \delta^h_k \nabla_i \Phi_{j]} - \delta^h_j \nabla_i \Phi_{k]} + g_{ij} g^{hm} \nabla_m \Phi_{k]} - g_{ik} g^{hm} \nabla_m \Phi_{j]}\right\} \quad (30)$$

where $P = \left\{\frac{\alpha n+\beta}{n(n-1)} + \frac{2b\beta}{an}\right\}$ and $Q = \frac{-b}{a}\beta$ are scalars.

Ricci tensor R_{ij} is obtained as

$$R_{ij} = \{P(n-1) + Q\}g_{ij} + Q(n-2)A_i A_j - \frac{n(n-2)b}{a}\nabla_i \Phi_{j]} \quad (31)$$

by contracting on the indices h and k in (30) and the scalar curvature is found in the form of

$$R = \{P(n-1) + Q\}n + Q(n-2) \quad (32)$$

by transvecting (31) by g^{ij}.

Using (10), (30), (31) and (32) in (11), it is obtained that

$$C^h_{ijk} = 0 \quad (33)$$

leading us to following:

Corollary 1. *Quasi conformally flat quasi Einstein-Weyl manifold* $(QEW)_n$ $(n > 3)$ *is conformally flat.*

3. Some necessary and sufficient conditions on quasi conformally flat quasi Einstein-Weyl manifolds

The concept of a space of quasi constant curvature was firstly introduced by Chen and Yano [15]. Similarly, we can define a Weyl manifold of quasi constant curvature as follows:

Definition 2. *A Weyl manifold* $\left(M, g_{ij}, \Phi_k\right)$ $(n > 3)$ *is said to be of quasi constant curvature if it is conformally flat and its curvature tensor* R^h_{ijk} *of type* (1,3) *is in the form of*

$$R^h_{ijk} = U\left(\delta^h_k g_{ij} - \delta^h_j g_{ik}\right) + V\left\{\delta^h_k A_i A_j - \delta^h_j A_i A_k + g_{ij} g^{hm} A_m A_k - g_{ik} g^{hm} A_m A_j\right\}, \quad (34)$$

where U *and* V *are scalars with* $V \neq 0$ *and* A_i *is a covariant vector.*

On the other hand, Amur and Maralabhavi [16] proved that a quasi conformally flat Riemannian manifold is either conformally flat or Einstein. So, a quasi conformally flat quasi Einstein manifold, which is not Einstein, is conformally flat and its curvature tensor satisfies the condition in (32) with $a \neq 0$ and $b \neq 0$. Therefore, a quasi conformally flat quasi Einstein manifold with $a \neq 0$ and $b \neq 0$ is of quasi constant curvature.

However, the situation is more complicated for quasi conformally flat $(QEW)_n$. Because although quasi conformally flat $(QEW)_n$ is conformally flat, it does not meet the requirement in (34) automatically. Therefore, a quasi conformally flat $(QEW)_n$ will be of quasi constant curvature under special conditions.

Suppose that quasi conformally flat $(QEW)_n$ $(n > 4)$ be of quasi constant curvature with the same definition in (2). Since $a \neq 0$ and $b \neq 0$, from (29),

$$2\delta_i^h \nabla_k \Phi_{j]} + \delta_k^h \nabla_i \Phi_{j]} - \delta_j^h \nabla_i \Phi_{k]} + g_{ij} g^{hm} \nabla_m \Phi_{k]} - g_{ik} g^{hm} \nabla_m \Phi_{j]} = 0 \tag{35}$$

is obtained. By transvecting (35) by g^{ij},

$$(n-4) g^{hi} \nabla_i \Phi_{k]} = 0 \tag{36}$$

and transvecting one more time by g_{hj} with the assumption of $n > 4$, it is found that

$$\nabla_j \Phi_{k]} = 0 \tag{37}$$

which means that the covariant derivative $\Phi_{k,j}$ is symmetric.

Conversely, let the covariant derivative $\Phi_{k,j}$ be symmetric in a quasi conformally flat $(QEW)_n$ $(n > 4)$. If (37) is substituted in (30), then (34) is obtained. Hence we get the following:

Theorem 2. *A necessary and sufficient condition for a quasi conformally flat quasi Einstein-Weyl manifold $(QEW)_n$ $(n > 4)$ to be of quasi constant curvature is that the covariant derivative $\Phi_{k,j}$ is symmetric.*

Now, let us consider in which cases the covariant derivative $\Phi_{k,j}$ is symmetric in a quasi conformally flat $(QEW)_n$, remembering that Φ_k is different from zero or non-gradient. So, let us give the definitions of some special vector fields in the Weyl manifold $\left(M, g_{ij}, \Phi_k\right)$:

Definition 3. *A vector field ξ in the Weyl manifold $\left(M, g_{ij}, \Phi_k\right)$ is called* ***torse-forming*** *if it satisfies the condition $\nabla_X \xi = \rho X + \lambda(X)\xi$, where $\xi \in \chi(M)$, $\lambda(X)$ is a linear form and ρ is a function. In the local coordinates, it is expressed by $\nabla_i \xi^h = \rho \delta_i^h + \xi^h \lambda_i$, where δ_i^h is the Kronecker symbol, ξ^h and λ_i are the components of ξ and λ. A torse-forming vector field ξ is called* ***concircular*** *if $\nabla_i \xi_j = \rho g_{ij}$ with $\xi_j = g_{hj} \xi^h$.*

Definition 4. *A vector field ϕ in the Weyl manifold $\left(M, g_{ij}, \Phi_k\right)$ is called* ***ϕ (Ric) vector field*** *if it satisfies $\nabla \phi = \mu Ric$, where μ is a constant and Ric is the Ricci tensor. In local coordinates, it is expressed by $\nabla_i \phi_j = \mu R_{ij}$, where ϕ_i and R_{ij} are the components of ϕ and Ric.*

Definition 5. *The components ϕ^i of a vector field ϕ in the Weyl manifold $\left(M, g_{ij}, \Phi_k\right)$ is defined as* ***parallel*** *if $\phi^i_{,j} = 0$ and is defined* ***concurrent*** *if $\phi^i_{,j} = c\delta^i_j$, where c is a constant.*

When we apply the above definitions to parallel, concurrent and concircular complementary vector field Φ_k in a quasi conformally flat quasi Einstein-Weyl manifold $(QEW)_n$, the covariant derivatives $\Phi_{k,j}$ of these vector fields are

$$\Phi_{k,j} = 2\Phi_k \Phi_j, \quad \Phi_{k,j} = 2\Phi_k \Phi_j + c g_{kj}, \quad \Phi_{k,j} = \rho g_{kj}, \tag{38}$$

respectively.

Now, let us consider Φ_k as a ϕ (Ric) vector field. From Definition 3.4 and (10),

$$(1-\mu n)\nabla_j \Phi_{k]} = 0. \tag{39}$$

Finally, let us write the covariant derivative $\Phi_{k,j}$ for a torse forming vector field Φ_k defined by $\nabla_i \Phi^h = \rho \delta_i^h + \Phi^h A_i$, where A_i is the generator of $(QEW)_n$. By using Definition 3.3, we have

$$\Phi_{k,j} = 2\Phi_j \Phi_k + \rho g_{kj} + A_j \Phi_k. \tag{40}$$

By means of (38), (39) and (40), we can express the following:

Corollary 2. *A quasi conformally flat quasi Einstein-Weyl manifold $(QEW)_n$ $(n > 4)$ is of quasi constant curvature if the complementary vector field Φ_k satisfies any one of the following:*

1. Φ_k *is a parallel, concurrent or concircular vector field,*

2. Φ_k *is a ϕ (Ric) vector field with $\mu \neq \frac{1}{n}$,*

3. Φ_k *is a torse forming vector field defined by $\nabla_i \Phi^h = \rho \delta_i^h + \Phi^h A_i$, where $A_j \Phi_k - A_k \Phi_j = 0$.*

Now, we seek a necessary and sufficient condition for a quasi conformally flat quasi Einstein-Weyl manifold $(QEW)_n$ to be recurrent. So, firstly, let us define the concept of recurrency in the quasi Einstein-Weyl manifolds by analogy to A.G. Walker's definition [17]:

Definition 6. *A non-flat Weyl manifold $\left(M, g_{ij}, \Phi_k\right)$ is called recurrent if there exists a non-zero covariant vector μ_l such that*

$$R^h_{ijk,l} - \mu_l R^h_{ijk} = 0. \tag{41}$$

Suppose that quasi conformally flat quasi Einstein-Weyl manifold $(QEW)_n$ $(n > 3)$, whose associated scalars α and β satisfy

$$\frac{\alpha_{,l}}{\alpha} = \frac{\beta_{,l}}{\beta} = \mu_l - 2\Phi_l, \tag{42}$$

is recurrent. From (39), it follows that

$$R_{ij,l} - \mu_l R_{ij} = 0 \tag{43}$$

by contracting on the indices h and k in (41) and transvecting (43) by g^{ij} gives us

$$R_{,\,l} - (\mu_l - 2\Phi_l)R = 0 \tag{44}$$

by means of (20).

By substituting (43) in (23), it is obtained that

$$n\{\alpha_{,l}-(\mu_l-2\Phi_l)\alpha\}+\{\beta_{,l}-(\mu_l-2\Phi_l)\beta\}=0 \tag{45}$$

which is satisfied by associated scalars in the above hypothesis.
If (42) is substituted in (19) and transvecting by A^i

$$0=\beta(A_{j,l}-\Phi_l A_j)+\frac{n}{2}\left[(\Phi_{j,il}-\Phi_{i,jl})-\mu_l(\Phi_{j,i}-\Phi_{i,j})\right]A^i \tag{46}$$

is obtained.
The conditions

$$A_{j,l}-\Phi_l A_j=0\Leftrightarrow A_{j,l}=\Phi_l A_j \tag{47}$$

and

$$\left[(\Phi_{j,il}-\Phi_{i,jl})-\mu_l(\Phi_{j,i}-\Phi_{i,j})\right]A^i=0 \tag{48}$$

are satisfied in order to provide (46) since $\beta\neq 0$ and $n>3$ for a quasi conformally flat quasi Einstein-Weyl manifold $(QEW)_n$.
If (47) is satisfied, then

$$A^j_{,l}=-\Phi_l A^j, \tag{49}$$

where $A^j=g^{jh}A_h$.
Let us first compute first and second covariant derivatives of the complementary vector Φ_j defined by (25) by considering (21), (47) and (48):

$$\Phi_{j,i}=A^k A_{k,i}-\Phi_j\Phi_i, \tag{50}$$

$$\Phi_{j,il}=2\Phi_j\Phi_i\Phi_l-A^k(A_{k,il}\Phi_j+A_{k,jl}\Phi_i+A_{k,ji}\Phi_l-A_{k,jil}). \tag{51}$$

By using (50) and (51), the expressions $(\Phi_{j,il}-\Phi_{i,jl})A^i$ and $\mu_l(\Phi_{j,i}-\Phi_{i,j})A^i$ in (48) can be written as

$$(\Phi_{j,il}-\Phi_{i,jl})A^i=A^iA^k\left[\Phi_l(A_{k,ij}-A_{k,ji})+(A_{k,jil}-A_{k,ijl})\right], \tag{52}$$

$$\mu_l(\Phi_{j,i}-\Phi_{i,j})A^i=A^iA^k\left[\mu_l(A_{k,ji}-A_{k,ij})\right]. \tag{53}$$

If (52) and (53) are substituted in (48), then

$$A^iA^k\left[(A_{k,jil}-A_{k,ijl})-(\mu_l+\Phi_l)(A_{k,ji}-A_{k,ij})\right]=0 \tag{54}$$

is obtained. Since A^i and A^k's are linearly independent,

$$A_{k,jil}=(\mu_l+\Phi_l)A_{k,ji}. \tag{55}$$

Conversely, let (47) and (55) be satisfied in a quasi conformally flat quasi Einstein-Weyl manifold $(QEW)_n$ whose associated scalars α and β satisfying (42).

From (30),

$$\begin{aligned}
R^h_{ijk,l} - \mu_l R^h_{ijk} = {} & \{P_{,l} - (\mu_l - 2\Phi_l)P\}\left(\delta^h_k g_{ij} - \delta^h_j g_{ik}\right) \\
& + (Q_{,l} - \mu_l Q)\left\{\delta^h_k A_i A_j - \delta^h_j A_i A_k + g_{ij} g^{hm} A_m A_k - g_{ik} g^{hm} A_m A_j\right\} \\
& + Q\Big\{\delta^h_k (A_{i,l} A_j + A_i A_{j,l}) - \delta^h_j (A_{i,l} A_k + A_i A_{k,l}) \\
& + g_{ij} g^{hm} (A_{m,l} A_k + A_m A_{k,l}) - g_{ik} g^{hm} (A_{m,l} A_j + A_m A_{j,l})\Big\} \\
& - \frac{(n-2)b}{a}\Big\{\delta^h_i \left[(\Phi_{j,kl} - \Phi_{k,jl}) - \mu_l (\Phi_{j,k} - \Phi_{k,j})\right] \\
& + \frac{1}{2}\delta^h_k \left[(\Phi_{j,il} - \Phi_{i,jl}) - \mu_l (\Phi_{j,i} - \Phi_{i,j})\right] \\
& - \frac{1}{2}\delta^h_j \left[(\Phi_{k,il} - \Phi_{i,kl}) - \mu_l (\Phi_{k,i} - \Phi_{i,k})\right] \\
& + \frac{1}{2} g_{ij} g^{hm} \left[(\Phi_{k,ml} - \Phi_{m,kl}) - \mu_l (\Phi_{k,m} - \Phi_{m,k})\right] \\
& - \frac{1}{2} g_{ik} g^{hm} \left[(\Phi_{j,ml} - \Phi_{m,jl}) - \mu_l (\Phi_{j,m} - \Phi_{m,j})\right]\Big\}
\end{aligned} \tag{56}$$

If (42), (47) and (55) are written in (56), then (41) is obtained. Hence we can state the following:

Theorem 3. *A necessary and sufficient condition for a quasi conformally flat quasi Einstein-Weyl manifold* $(QEW)_n$ $(n > 3)$ *whose recurrent scalars* α *and* β *having the same recurrency vector* $\mu_l - 2\Phi_l$ *to be recurrent is that the equations* $A_{k,j} = \Phi_j A_k$ *and* $A_{k,jil} = (\mu_l + \Phi_l) A_{k,ji}$ *are satisfied.*

Let us dedicate the last part of this section to the concept of semi-symmetricness in a quasi conformally flat quasi Einstein-Weyl manifold $(QEW)_n$. Firstly, let us define semi-symmetric $(QEW)_n$ similar to the definition which is made by Szabo for Riemannian manifolds [18] as follows:

Definition 7. *A non-flat Weyl manifold* $\left(M, g_{ij}, \Phi_k\right)$ *is called semi-symmetric if its curvature tensor* R^h_{ijk} *of type* (1,3) *satisfies the condition*

$$R^h_{ijk,lm} - R^h_{ijk,ml} = 0. \tag{57}$$

It follows that

$$R_{ij,lm} - R_{ij,ml} = 0 \tag{58}$$

by contracting on the indices h and k in (57).

Let us suppose that a quasi conformally flat quasi Einstein-Weyl manifold $(QEW)_n$ $(n > 3)$ with $\alpha n \neq b\beta(n-2)$ is semi-symmetric. From (17) and (58),

$$\begin{aligned}
0 = {} & \beta\left[(A_{i,lm} - A_{i,ml})A_j + A_i\left(A_{j,lm} - A_{j,ml}\right)\right] + 2\alpha g_{ij}(\Phi_{l,m} - \Phi_{m,l}) \\
& + \frac{n}{2}\left[\left(\Phi_{j,ilm} - \Phi_{i,jlm}\right) - \left(\Phi_{j,iml} - \Phi_{i,jml}\right)\right].
\end{aligned} \tag{59}$$

With the aid of the Ricci identity given as

$$v_{i,jk} - v_{i,kj} = v_h R^h_{ijk}, \tag{60}$$

where v_i's are the components of a covariant vector, it is obtained that

$$\begin{aligned} 0 = \; & \beta\left[\left(A_h R^h_{ilm}\right)A_j + A_i\left(A_h R^h_{jlm}\right)\right] + 2\alpha g_{ij}(\Phi_{l,m} - \Phi_{m,l}) \\ & + \frac{n}{2}\left[\left(\Phi_{j,ilm} - \Phi_{i,jlm}\right) - \left(\Phi_{j,iml} - \Phi_{i,jml}\right)\right]. \end{aligned} \tag{61}$$

If (61) is transvected by g^{ij}, it is found that

$$0 = 2\beta\left(A_h R^h_{ilm}\right)A^i + 2\alpha n(\Phi_{l,m} - \Phi_{m,l}). \tag{62}$$

Substituting the following equation, resulted from (48),

$$\left(A_h R^h_{ilm}\right)A^i = -\frac{(n-2)b}{a}(\Phi_{l,m} - \Phi_{m,l}) \tag{63}$$

which is valid in a quasi conformally flat quasi Einstein-Weyl manifold $(QEW)_n$ in (62) gives us

$$(\Phi_{l,m} - \Phi_{m,l})\left(\alpha n - \beta\frac{(n-2)b}{a}\right) = 0. \tag{64}$$

Because of the restriction on α and β,

$$\Phi_{l,m} - \Phi_{m,l} = 0. \tag{65}$$

If we form the difference $\Phi_{l,m} - \Phi_{m,l}$ after taking covariant derivative of (25) with repect to x^m, we have

$$\Phi_{l,m} - \Phi_{m,l} = A^i(A_{i,lm} - A_{i,ml}) + A_{i,l}A^i_{\ ,m} - A_{i,m}A^i_{\ ,l}. \tag{66}$$

If firstly rearranging the first term on the right hand side of the equation in (66) with the help of (60) and then using (63) and (64) in the resulting equation gives

$$A_{i,l}A^i_{\ ,m} - A_{i,m}A^i_{\ ,l} = 0 \tag{67}$$

or equivalently

$$A_{i,l}A_{j,m} - A_{j,l}A_{i,m} = 0. \tag{68}$$

Conversely, let us assume that the generator A_i of a quasi conformally flat quasi Einstein-Weyl manifold $(QEW)_n$ $(n > 3)$ satisfies the condition (67) or equivalently (68). If (67) is substituted in (66), then (65) is satisfied by means of (60). In this case,

$$A^i(A_{i,lm} - A_{i,ml}) = A^i\left(A_h R^h_{ilm}\right) = 0 \tag{69}$$

which means that

$$A_{i,lm} - A_{i,ml} = 0 \tag{70}$$

since A^i's are linearly independent.

From (30),

$$\begin{aligned}
R^h_{ijk,ls} = {} & (P_{,ls} + 2\Phi_{l,s}P + 2\Phi_l P_{,s} + 2\Phi_s P_{,l} + 4\Phi_l\Phi_s P)\left(\delta^h_k g_{ij} - \delta^h_j g_{ik}\right) \\
& + Q_{,ls}\left\{\delta^h_k A_i A_j - \delta^h_j A_i A_k + g_{ij}g^{hm}A_m A_k - g_{ik}g^{hm}A_m A_j\right\} \\
& + Q_{,l}\left\{\delta^h_k (A_{i,s}A_j + A_i A_{j,s}) - \delta^h_j (A_{i,s}A_k + A_i A_{k,s}) + g_{ij}g^{hm}(A_{m,s}A_k + A_m A_{k,s})\right\} \\
& - g_{ik}g^{hm}(A_{m,s}A_j + A_m A_{j,s}) \\
& + Q_{,s}\{\delta^h_k (A_{i,l}A_j + A_i A_{j,l}) - \delta^h_j (A_{i,l}A_k + A_i A_{k,l}) + g_{ij}g^{hm}(A_{m,l}A_k + A_m A_{k,l}) \\
& - g_{ik}g^{hm}(A_{m,l}A_j + A_m A_{j,l})\} + Q\{\delta^h_k (A_{i,ls}A_j + A_{i,l}A_{j,s} + A_{i,s}A_{j,l} + A_i A_{j,ls})\} \\
& - \delta^h_j (A_{i,ls}A_k + A_{i,l}A_{k,s} + A_{i,s}A_{k,l} + A_i A_{k,ls}) \\
& + g_{ij}g^{hm}(A_{m,ls}A_k + A_{m,l}A_{k,s} + A_{m,s}A_{k,l} + A_m A_{k,ls}) \\
& - g_{ik}g^{hm}(A_{m,ls}A_j + A_{m,l}A_{j,s} + A_{m,s}A_{j,l} + A_m A_{j,ls})\}
\end{aligned} \tag{71}$$

is obtained. If necessary simplifications are made in the difference $R^h_{ijk,ls} - R^h_{ijk,sl}$ which is formed by means of (66), then it is found that

$$\begin{aligned}
R^h_{ijk,ls} - R^h_{ijk,sl} = {} & Q\{\delta^h_k[(A_{i,ls} - A_{i,sl})A_j + A_i(A_{j,ls} - A_{j,sl})] \\
& - \delta^h_j[(A_{i,ls} - A_{i,sl})A_k + A_i(A_{k,ls} - A_{k,sl})] \\
& + g_{ij}g^{hm}[(A_{m,ls} - A_{m,sl})A_k + A_m(A_{k,ls} - A_{k,sl})] \\
& - g_{ik}g^{hm}[(A_{m,ls} - A_{m,sl})A_j + A_m(A_{j,ls} - A_{j,sl})]\}
\end{aligned} \tag{72}$$

If (70) is written in (72), then we have

$$R^h_{ijk,ls} - R^h_{ijk,sl} = 0 \tag{73}$$

which tells us that quasi conformally flat quasi Einstein-Weyl manifold $(QEW)_n$ is semi-symmetric. Therefore we can express the following:

Theorem 4. *A necessary and sufficient condition for a quasi conformally flat quasi Einstein-Weyl manifold* $(QEW)_n$ $(n > 3)$ *with* $\alpha an \neq \beta b(n-2)$ *to be semi-symmetric is that the equation* $A_{i,l}A^i_{,m} - A_{i,m}A^i_{,l} = 0$ *is satisfied.*

In the last part of this section, let us take a look at the relationships between to be of quasi constant curvature and to be semi-symmetric in a quasi conformally flat quasi Einstein-Weyl manifold $(QEW)_n$.

If we combine Theorem 1 with Theorem 3 we get the following:

Corollary 3. *A necessary and sufficient condition for a quasi conformally flat quasi Einstein-Weyl manifold* $(QEW)_n$ $(n > 4)$ *with* $\alpha an \neq \beta b(n-2)$ *to be semi-symmetric is that the manifold is of quasi constant curvature.*

Now, we will examine two special cases of the generator A^i of a quasi conformally flat quasi Einstein-Weyl manifold $(QEW)_n$:

I.case: For a parallel generator A^i; since

$$A_{i,l} = 2A_i\Phi_l \text{ and} A_{i,lm} - A_{i,ml} = 2A_i(\Phi_{l,m} - \Phi_{m,l}) \tag{74}$$

from Definition 5, it is clear that a quasi conformally flat quasi Einstein-Weyl manifold $(QEW)_n$ $(n > 4)$, which the generator A^i is parallel, is automatically semi-symmetric. If Definition 5 and the equations in (74) are used in (66), then (65) is obtained. This means that the manifold is of quasi constant curvature by means of Theorem 2.

Conversely, let us assume that quasi conformally flat quasi Einstein-Weyl manifold $(QEW)_n$ $(n > 4)$, which the generator A^i is parallel, is of quasi constant curvature. In this case, if (70), which is implied by (64), is substituted in (72), then (73) is achieved which means that the manifold is semi-symmetric. Hence we can state the following:

Theorem 5. *A necessary and sufficient condition for a quasi conformally flat quasi Einstein-Weyl manifold $(QEW)_n$ $(n > 4)$ which the generator A^i is parallel to be semi-symmetric is that the manifold is of quasi constant curvature.*

II.case: Let us consider a quasi conformally flat quasi Einstein-Weyl manifold $(QEW)_n$ $(n > 4)$ with $2\beta + \alpha n \neq 0$, which the generator A^i is concurrent, is semi-symmetric. From Definition 5, it follows that

$$A_{i,l} = 2A_i\Phi_l + cg_{il}, \qquad A_{i,lm} - A_{i,ml} = 2A_i(\Phi_{l,m} - \Phi_{m,l}),$$
$$A_{i,l}A^i_{,m} - A_{i,m}A^i_{,l} = c(A_{m,l} - A_{l,m}) \ . \tag{75}$$

If Definition 5 and the equations in (75) are used in (66), then

$$\Phi_{l,m} - \Phi_{m,l} = c(A_{l,m} - A_{m,l}). \tag{76}$$

Using (75) and (76) in (62) gives

$$c(4\beta + 2\alpha n)(A_{l,m} - A_{m,l}) = 0. \tag{77}$$

Because of the assumption on α and β, $A_{l,m} - A_{m,l} = 0$ and therefore $\Phi_{l,m} - \Phi_{m,l} = 0$ by (76) which tells us that the manifold is of quasi constant curvature.

Conversely, a quasi conformally flat quasi Einstein-Weyl manifold $(QEW)_n$ $(n > 4)$, which the generator A^i is concurrent, is of quasi constant curvature. Then, by Theorem 1, $\Phi_{l,m} - \Phi_{m,l} = 0$ which is equivalent to $A_{l,m} - A_{m,l} = 0$ by (76). If the last equation is substituted in (72), then (73) is obtained which means that the manifold is semi-symmetric. Hence we can state the following:

Theorem 6. *A necessary and sufficient condition for a quasi conformally flat quasi Einstein-Weyl manifold $(QEW)_n$ $(n > 4)$ with $2\beta + \alpha n \neq 0$ which the generator A^i is concurrent to be semi-symmetric is that the manifold is of quasi constant curvature.*

Acknowledgements

The author is grateful to the referee for his/her valuable comments and suggestions for the improvement of the book chapter.

Author details

Fusun Nurcan
Department of Mathematics, Marmara University, Istanbul, Turkey

*Address all correspondence to: fnurcan622@gmail.com

References

[1] Weyl H. Gravitation und Elektrizität. Berlin: S.B. Press, Akad.Wiss; 1918

[2] Cartan E. Sur une classe d'espaces de Weyl. Annales Scientifiques de l'École Normale Supérieure. 1943;**60**:1-16

[3] Jones PE, Tod KP. Minitwistor and Einstein Weyl spaces. Classical and Quantum Gravity. 1985;**2**:567-577

[4] Norden A. Affinely Connected Spaces. Moscow: GRMFL; 1976

[5] Hlavaty V. Theorie d'immersion d'une W_m dans W_n. Annales Polonici Mathematici. 1949;**21**:196-206

[6] Miron R. Mouvements conformes dans les espaces W_n and N_n. Tensor. 1968;**19**:33-41

[7] Ozdeger A, Senturk Z. Generalized Circles in Weyl Spaces and theirconformal mapping. Universitatis Debreceniensis. 2002;**60**:75-87

[8] Yano K, Sawaki S. Riemannian Manifolds admitting a conformal transformation group. Journal of Differential Geometry. 1968;**2**:161-184

[9] Nurcan F. Quasi conformally symmetric Weyl Manifolds. International Journal of Pure and Applied Mathematics. 2015;**3**:415-425

[10] Chaki MC, Maity RK. On quasi Einstein manifolds. Universitatis Debreceniensis. 2000;**57**:297-306. DOI: 10.1016/s0014-5793(01)03293-8

[11] De UC, Sengupta J, Saha D. Conformally flat quasi Einstein spaces. Kyungpook National University. 2006; **46**:417-423

[12] Shaikh AA, Patra A. On quasi-conformally flat quasi-Einstein spaces. Differential Geometry-Dynamical Systems. 2010;**12**:201-212

[13] Shaikh AA, Patra A. On quasi-conformally flat quasi-Einstein spaces with recurrent curvature. International Journal of Mathematical Analysis. 2011; **12**:559-568

[14] Gül İ, Canfes EÖ. On quasi Einstein-Weyl Manifolds. International Journal of Geometric Methods in Modern Physics. 2017;**1**:1750-1822

[15] Chen BY, Yano K. Hypersurfaces of a conformally flat space. Tensor. 1972;**26**: 318-322

[16] Amur K, Maralabhavi YB. On quasi conformally flat spaces. Tensor. 1977;**31**: 94-108

[17] Walker AG. On Ruse's spaces of recurrent curvature. Proceedings of the London Mathematical Society. 1951;**52**: 36-64

[18] Szabo ZI. Structure theorems on Riemannian spaces satisfying R(X,Y).R= 0. Journal of Differential Geometry. 1982;**17**:531-582

Chapter 3

An Introduction to the Generalized Gauss-Bonnet-Chern Theorem

Paul Bracken

Abstract

This work studies the mathematical structures which are relevant to differentiable manifolds needed to prove the Gauss-Bonnet-Chern theorem. These structures include de Rham cohomology vector spaces of the manifold, characteristic classes such as the Euler class, pfaffians, and some fiber bundles with useful properties. The paper presents a unified approach that makes use of fiber bundles and leads to a non-computational proof of the Gauss-Bonnet-Chern Theorem. It is indicated how it can be generalized to manifolds with boundary.

Keywords: manifold, Euler characteristic, bundle, fiber, projection, universal

1. Introduction

One of the great achievements of differential geometry is the Gauss-Bonnet theorem. In its original form, the theorem is a statement about surfaces which connect their geometry in the sense of curvature to the underlying topology of the space, in the sense of the Euler characteristic [1–5]. The most elementary case of the theorem states that the sum of the angles of a triangle in the plane is π radians. If the surface is deformed, the Euler characteristic does not vary as it is a topological invariant, while the curvature at certain points does change [6–8]. The theorem states that the total integral of the curvature remains the same, no matter how the deformation is performed. If there is a sphere with a ding, its total curvature is 4π since its Euler characteristic is two. This is the case no matter how big or deep the actual deformation is. A torus has Euler characteristic zero, so its total curvature must also be zero. If the torus carries the usual Riemannian metric from its embedding in $\mathbb{R}^3$, then the inside has negative Gaussian curvature, and the outside has positive Gaussian curvature, so the total curvature is zero. It is not possible to specify a Riemannian metric on the torus which has everywhere positive or everywhere negative Gaussian curvature. Manifolds M have dimension n unless stated otherwise [9–11]. There are many applications of this theorem in both mathematics and mathematical physics such as in gravity [12–14], string theory [15] and even in the study of Ricci flow [16].

Although the curvature K is defined intrinsically in terms of the metric on the manifold M. It can also be defined for $n = 2$ extrinsically when the metric on M is induced by an embedding $M \subset \mathbb{R}^3$. In fact, it $\nu : M \to S^2$ is the normal map and da is the volume element on S^2, then $K\,d\sigma = \nu^*\,da$ so that

$$\int_M K d\sigma = \int_M \nu^* (da) = \deg(\nu) \cdot \int_{S^2} da = 4\pi \cdot \deg(\nu). \tag{1}$$

Without bringing in differential geometric considerations, it is seen to be the case that $\deg(\nu) = (1/2)\chi(M)$, where $\chi(M)$ is the Euler characteristic of M. Using this fact in (1), the Gauss-Bonnet theorem for a compact oriented surface M, the first version of the theorem is obtained for the case in which the metric on M arises by means of an embedding in $\mathbb{R}^3$

$$\int_M K d\sigma = 2\pi\chi(M). \tag{2}$$

It is the intention here to state and prove a general version of the theorem which applies to manifolds of even dimension, so a surface with $n = 2$ is a special case. An intrinsic proof of the theorem was obtained by Chern 1944. The kind of argument outlined above was used by Hopf in developing the first generalization of the theorem. To outline the basic idea, consider a compact surface $M^n \subset \mathbb{R}^{n+1}$ when n is even. If $d\mu_g$ is the volume form on the manifold and ds_n denotes the volume element S^n, then

$$\int_{M^n} K_n d\mu_g = \int_{M^n} \nu^* ds_n = \mathrm{vol}(S^n) \cdot \deg(\nu) = \frac{1}{2} \mathrm{vol}(S^n) \cdot \chi(M^n). \tag{3}$$

This can be extended to any compact oriented Riemannian n-manifold (M^n, g) which has even dimension, where K_n in a coordinate system is given by

$$K_n = \frac{1}{2^{n/2} n!} \sum_{\substack{i_1, \dots, i_n \\ j_1, \dots, j_n}} R_{i_1 i_2 j_1 j_2} \cdots R_{i_{n-1} i_n j_{n-1} j_n} \frac{1}{\sqrt{g}} \varepsilon^{i_1 \dots i_n} \cdot \frac{1}{\sqrt{g}} \varepsilon^{j_1 \dots j_n}. \tag{4}$$

The $\sqrt{g}$ (4) is the square root of the determinant of the metric. With K_n given by (4), and μ_g the volume form on the manifold, we are then led to conjecture that

$$\int_{M^n} K_n d\mu_g = \frac{1}{2} \mathrm{vol}(S^n) \cdot \chi(M^n), \tag{5}$$

where M is a compact, oriented Riemannian manifold with n even.

It is the objective to look at and study some of the ensuing developments which have led to a much deeper understanding of the foundations which underlie this theorem. It will be seen that this development leads to a completely non-computational proof of this deep theorem.

2. Characteristic classes

When an oriented n-dimensional manifold $\left(M, d\mu_g\right)$ is compact and closed, with $d\mu_g$ is the volume form and μ the orientation of M, so every form has compact support, Stokes theorem leads to the important theorem. Let η be any $(n-1)$-form then

$$\int_M d\eta = \int_{\partial M} \eta. \tag{6}$$

Therefore an n-form ω on M, which is not exact, even though it must be closed as all n-forms on M are zero, can be found simply by locating an ω such that

$$\int_M \omega \neq 0. \tag{7}$$

Such a form always exists, as it is known there is a form ω such that for $v_1, \dots, v_n \in M_p$, $\omega(v_1, \dots, v_n) > 0$ if $[v_1, \dots, v_n] = \mu_p$. If $c : [0,1]^n \to (M, \mu)$ preserves orientation, $c^* \omega$ on $[0,1]^n$ is $g\, dx^1 \wedge \cdots \wedge dx^n$ for some $g > 0$ on $[0,1]^n$, hence $\int_c \omega > 0$. This observation leads to this theorem. A smooth, oriented manifold is not smoothly contractible to a point. In fact, it is the shape of M not the size which determines whether or not every closed form on M is exact. More information about the shape of M can be obtained by analyzing more closely the extent to which closed forms are not exact. So how many non-exact n-forms are there on a compact oriented n-manifold If ω is not exact, the same holds for $\omega + d\eta$ for η any $(n-1)$-form η. Thus it is necessary to regard ω and $\omega + d\eta$ as equivalent. This suggests an equivalence relation and directs one to think of this in terms of quotient spaces.

For each k, $Z^k(M)$ denotes all closed k-forms on M and it is a vector space. The space $B^k(M)$ of all exact k-forms is a subspace since $d^2 = 0$. The quotient space is called the k-dimensional de Rham cohomology vector space of M and is defined to be

$$H^k(M) = Z^k(M)/B^k(M). \tag{8}$$

The theorem of de Rham states that the vector space is isomorphic to a vector space defined just in terms of the topology of M called the k-dimensional cohomology group of M with real coefficients.

An element of $H^k(M)$ is an equivalence class $[\omega]$ of a closed form ω such that closed forms ω_1 and ω_2 are equivalent if and only if the difference is exact. In terms of these vector spaces, the Poincaré lemma gives $H^k(\mathbb{R}^n) = 0$, the vector space consisting of just the zero vector if $k > 0$, or $H^k(M) = 0$ if M is contractible and $k > 0$. To compute $H^0(M)$ note $B^0(M) = 0$ as there are no non-zero exact 0-forms as there are no non-zero minus one forms. Thus $H^0(M)$ is the same as the vector space of all C^∞ functions $f : M \to \mathbb{R}$ with $df = 0$. If M is connected, this condition implies f is constant so $H^0(M) \equiv \mathbb{R}$ and its dimension is the number of components of M.

The de Rham cohomolgy vector spaces with compact support $H_c^k(M)$ are defined similarly to (8), that is, $H_c^k(M) = Z_c^k(M)/B_c^k(M)$, where $Z_c^k(M)$ is the vector space of all closed k-forms with compact support and $B_c^k(M)$ all k-forms $d\eta$ where η is a k-form with compact support. If M is compact $H_c^k(M) = H^k(M)$.

Theorem 2.1. (The Poincaré-Duality Theorem) If M is a connected, oriented n-manfold of finite type, then the map

$$\Pi : H^k(M) \to H_c^{n-k}(M) \tag{9}$$

is an isomorphism for all k. □

This theorem eventually motivates the introduction of the Euler characteristic for any smooth connected oriented manifold M. Consider then a smooth k-dimensional vector bundle $\xi = \pi : E \to M$ over M. Orientations μ for M, and ν for ξ give an orientation $\mu \oplus \nu$ for the $(n+k)$-manifold E, since E is locally a product. Let $\{U_1, \ldots, U_r\}$ be a cover of M by geodesically convex sets so small that each bundle ξ restricted to U_i is trivial. Then $\{\pi^{-1}(U_1), \ldots, \pi^{-1}(U_r)\}$ turns out to be a nice cover for E, so it is a manifold of finite type. For the section and projection maps s, π, $\pi \circ s = I$ on M and $s \circ \pi$ is smoothly homotopic to the identity of E, so the map $\pi^* : H^l(M) \to H^l(E)$ is an isomorphism for all l. The reason for mentioning (6) and Theorem 2.1 is that it shows there is a unique class $U \in H_c^k(E)$ such that

$$\pi^* : \mu \cup U = \mu \oplus \nu \subset H_c^{n+k}(E). \tag{10}$$

This class is called the Thom class of ξ.

A theorem states that if (M, μ) is a compact oriented, connected manifold $\xi = \pi : E \to M$ an oriented k-plane bundle over M orientation ν, the Thom class U is the unique element of $H_c^k(E)$ such that for all $p \in M$, and $j_p : F_p \to E$ the inclusion map, we have $j_p^* U = \nu_p$. This condition has the implication that $\int_{(F_p, \nu_p)} j_p \omega = 1$, where U is the class of closed form ω.

The Thom class U of $\xi = \pi : E \to M$ can now be used to determine an element of $H^k(M)$. Let $s : M \to E$ be any section. There is always one, any two are clearly homotopic. Define the Euler class $\chi(E) \subset H^k(M)$ of ξ by

$$\chi(\xi) = s^* U. \tag{11}$$

If ξ has a non-zero section $s : M \to E$ and $\omega \in C_c^k(M)$ represents U, a suitable multiple $c \cdot s$ of s takes M to the complement of support ω, so in this case, $\chi(\xi) = (c \cdot s)^* U = 0$.

The term Euler class is connected with the special case of the bundle TM which has sections which are vector fields on M. If X is a vector field on M having an isolated zero at some point p, $X(p) = 0$, but $X(q) \neq 0$ for $q \neq p$ in a neighborhood of p. An index of X at p can be defined. Suppose X is a vector field on an open set $U \subset \mathbb{R}^n$ with an isolated zero at $0 \in U$. Define $f_X : U \to \{0\} \to S^{n-1}$ by $f_X(p) = X(p)/|X(p)|$. If $i : S^{n-1} \to U$ is $i(p) = \varepsilon p$ mapping S^{n-1} into U, then the map $f_X \circ i : S^{n-1} \to S^{n-1}$ has a certain degree independent of ε for small ε, since maps $i_1, i_2 : S^{n-1} \to U$ correspinding to $\varepsilon_1, \varepsilon_2$ will be smoothly homotopic. This degree is called the index of X at 0. Consider a diffeomorphism $h : U \to V \subset \mathbb{R}^n$ with $h(0) = 0$, so $h_* X$ is the vector field on V such that $(h_* X)(y) = h_* \left(X_{h^{-1}(y)} \right)$. So 0 is an isolated zero of $h_* X$. It can be shown, if $h : U \to V \subset \mathbb{R}^n$ is a diffeomorphism with $h(0) = 0$ and X has an isolated zero at 0, the index of $h_* X$ at 0 equals the index of X at 0.

As a consequence of this, an index of a vector field on a mainifold can be defined. If X is a vector field on M, with isolated zero at $p \in M$, choose a coordinate system (x, U) such that $x(p) = p$ and define the index of X at p to be the index of $x_* X$ at 0.

Theorem 2.2. Let M be a compact, connected manifold with orientation μ, also an orientation for the tangent bundle $\xi = \pi : TM \to M$. Let $X : M \to TM$ be a vector field with only a finite number of zeros and let σ be the sum of indices of X at these zeros. Then

$$\chi(\xi) = \sigma \cdot \mu \in H^0(M). \tag{12}$$

□

3. Pfaffians

An intrinisic expression along with one in a coordinate system for the function K_n on a compact, oriented Riemannian manifold of even dimension has been given already. Another more important way of expressing K_n involves the curvature form Ω^i_j for a positively oriented orthonormal moving frame $X_1, \dots, X_n$ on M. In terms of these forms, the n-form $K_n \, d\mu_g$, the one to be integrated, can be written down. A sum over permutations such as

$$\sum_{\pi \in S_n} B\big(X_{\pi(1)}, \dots, X_{\pi(n)}\big), \tag{13}$$

can be written just as well as

$$\sum_{j_1, \cdots, j_n} \varepsilon^{j_1, \cdots, j_n} B\big(X_{j_1}, \dots, X_{j_n}\big). \tag{14}$$

Suppose this is the n-fold wedge product

$$\Omega^{i_1}_{i_2} \wedge \cdots \wedge \Omega^{i_{n-1}}_{i_n}. \tag{15}$$

Since the Ω^i_j are 2-forms, using the definition of wedge product,

$$\begin{aligned} \Omega^{i_1}_{i_2} \wedge \cdots \wedge \Omega^{i_{n-1}}_{i_n}(X_1, \dots, X_n) &= \frac{(2 + \cdots + 2)!}{2! \cdots 2!} \cdot \frac{1}{n!} \sum_{j_1, \cdots j_n} \varepsilon^{i_1 \cdots i_p} \, \Omega^{i_1}_{i_2}\big(X_{j_1}, X_{j_2}\big) \cdots \Omega^{i_{n-1}}_{i_n}\big(X_{j_{n-1}}, X_{j_n}\big) \\ &= \frac{1}{2^{n/2}} \sum_{j_1, \cdots j_n} \varepsilon^{j_1, \cdots, j_n} \big\langle R\big(X_{j_1}, X_{j_2}\big) X_{j_2}, X_{j_1} \big\rangle \cdots \big\langle R\big(X_{j_{n-1}}, X_{j_n}\big) X_{j_n}, X_{j_{n-1}} \big\rangle \\ &= \frac{1}{2^{n/2}} \sum_{j_1, \cdots, j_n} R_{i_1 i_2 j_1 j_2} \cdots R_{i_{n-1} i_n j_{n-1} j_n}. \end{aligned} \tag{16}$$

Comparing this to (4), it may be concluded that

$$K_n = \frac{1}{2^{n/2} n!} \sum_{i_1, \cdots, i_n} \varepsilon^{i_1, \cdots, i_n} 2^{n/2} \Omega^{i_1}_{i_2} \wedge \cdots \wedge \Omega^{i_{n-1}}_{i_n}(X_1, \cdots, X_n). \tag{17}$$

When (17) is multiplied by the volume form $d\mu_g$, it becomes

$$K_n \, d\mu_g = \frac{1}{n!} \sum_{i_1, \cdots, i_n} \Omega^{i_1}_{i_2} \wedge \cdots \wedge \Omega^{i_{n-1}}_{i_n}. \tag{18}$$

By (18) the form on the right does depend on the choice of the positively oriented orthonormal frame, $X_1, \dots, X_n$. There is a direct way to get this algebraically.

Suppose A is an $n \times n$ matrix $A = (a_{ij})$ with $n = 2m$ even. Define the Pfaffian, $\mathrm{Pf}(A)$ of A to be

$$\mathrm{Pf}(A) = \frac{1}{2^m m!} \sum_{i_1, \cdots, i_n} a_{i_1 i_2} \cdots a_{i_{n-1} i_n}. \tag{19}$$

Note that $\varepsilon^{i_1 \cdots i_n}$ does not change when any permutation of the pairs (i_{2l-1}, i_{2l}). For any set $S = \{(h_1, k_1), \cdots, (h_m, k_m)\}$ of pairs of integers between 1 and n, let us define

$$\varepsilon(S) = \varepsilon^{h_1 k_1 \cdots h_m k_m}. \tag{20}$$

It is not necessary to specify any ordering of pairs (h_i, k_i) in S. Also a permutation of the pairs (i_{2l-1}, i_{2l}) does not change the factor $a_{i_1 i_2} \cdots a_{i_{n-1} i_n}$. So for each P above define $a_S = a_{h_1 k_1} \cdots a_{h_m k_m}$. If P is the collection of all such S, we clearly have

$$\mathrm{Pf}(A) = \frac{1}{2^m} \sum_{S \in P} \varepsilon(S) a_S. \tag{21}$$

Theorem 3.1. Let $n = 2m$ then for all $n \times n$ matrices A and B,

$$\mathrm{Pf}\,(B^t A B) = (\det B) \cdot \mathrm{Pf}(A). \tag{22}$$

and B^t denotes the transpose. If $B \in SO(n)$ then

$$\mathrm{Pf}\left(B^{-1} A B\right) = \mathrm{Pf}(A). \tag{23}$$

Proof:

$$\begin{aligned} 2^m \cdot m!\ \mathrm{Pf}(B^t A B) &= \sum_{i_1 \cdots i_n} \varepsilon^{i_1 \cdots i_n} \sum_{j_1 \cdots j_n} \left(b_{j_1 i_1} a_{j_1 j_2} b_{j_2 i_2}\right) \cdots \left(b_{j_{m-1} i_{m-1}} a_{j_{m-1} j_n} b_{j_n i_n}\right) \\ = \sum_{j_1 \cdots j_n} \left[\sum_{i_1 \cdots i_n} \varepsilon^{j_1 \cdots j_n} b_{j_1 i_1} \cdots b_{j_n i_n} \right] a_{j_1 j_2} \cdots a_{j_{n-1} j_n} &= \sum_{j_1, \cdots, j_n} \varepsilon^{j_1, \cdots, j_n} \det(B) a_{j_1 j_2} \cdots a_{j_{n-1} j_n} \\ &= 2^m m! (\det B)\ \mathrm{Pf}(A). \end{aligned} \tag{24}$$

□

This theorem was stated for matrices of real numbers, but $\mathrm{Pf}(A)$ can be defined provided the entries of A are in some commutative algebra over $\mathbb{R}$.

Consider again a positively oriented orthogonal moving frame $X_1, \ldots, X_n$ on M, with curvature forms Ω^i_j. For each $p \in M$, the direct sum $A = \mathbb{R} \oplus \Omega^2(M_p) \oplus \Omega^2(M_p) \oplus \cdots$ is a commutative algebra over $\mathbb{R}$ under the operation $\wedge$. So one can consider $\mathrm{Pf}(\Omega(p))$, where $\Omega(p)$ is an $n \times n$ matrix of connection 2-forms at p

$$\mathrm{Pf}(\Omega_p) = \frac{1}{2^m m!} \sum_{i_1, \cdots, i_n} \varepsilon^{i_1, \cdots, i_n} \Omega^{i_1}_{i_1} \wedge \cdots \wedge \Omega^{i_{n-1}}_{i_n}(p). \tag{25}$$

If $X' = X \cdot a$ is another positively oriented orthonormal moving frame then $a(p) \in O(n)$ and the corresponding curvature forms satisfy $\Omega' = a^{-1} \Omega a$. Then Theorem 3.1 implies that

$$\mathrm{Pf}(\Omega'(p)) = \mathrm{Pf}\left(a^{-1}(p)\Omega a(p)\right) = \mathrm{Pf}(\Omega(p)) \tag{26}$$

so the form $\sum_{i_1,\cdots,i_n} \varepsilon^{i_1,\cdots,i_n} \Omega^{i_1}_{i_2} \wedge \cdots \wedge \Omega^{i_{n-1}}_{i_n}$ is well defined.

4. Bundles of paticular importance

Projective n-space $\mathbb{P}^n$ can be defined as the set of all pairs $(-p,p)$ for $p \in S^n \subset \mathbb{R}^{n+1}$ or the set of line through 0 in $\mathbb{R}^{n+1}$, since each lines intersects S^n through two antipodal points. A Grassmannian manifold $G_n(\mathbb{R}^n)$ is the set of all n-dimensional subspaces of $\mathbb{R}^N$ with $N > 0$. Over the Grassmannian manifold $G_n\left(\mathbb{R}^N\right)$, there is a natural n-dimensional bundle $\zeta^n(\mathbb{R}^n)$ constructed as follows. The total space of the bundle $E\left(\zeta^n\left(\mathbb{R}^N\right)\right)$ is the subset of $G_n\left(\mathbb{R}^N\right) \times \mathbb{R}^N$ consisting of all pairs

$$(W, w) \in G_n\left(\mathbb{R}^N\right) \times \mathbb{R}^M, \qquad w \in W. \tag{27}$$

The projection map which takes $E\left(\zeta^n\left(\mathbb{R}^N\right)\right) \to G_n\left(\mathbb{R}^N\right)$ is $\pi((W,w)) = W$. The fibre $\pi^{-1}(W)$ over W of $G_n\left(\mathbb{R}^N\right)$ will be W itself or, more explicitly, $\{(W,w) : w \in W\}$. A vector space structure is defined on $\pi^{-1}(W)$ by using the vector space structure on $W \subset \mathbb{R}^N$; if a is a scalar, then $(W,w_1) + (W,w_2) = (W, w_1 + w_2)$ and $a\,(W,w) = (W, aw)$. Also $\zeta^n\left(\mathbb{R}^N\right)$ satisfies the local triviality condition.

For $M > N$ there is a natural map $\alpha : G_n\left(\mathbb{R}^N\right) \to G_n\left(\mathbb{R}^M\right)$, as an n-dimensional subspace of $\mathbb{R}^N$ can be considered an n-dimensional subspace of $\mathbb{R}^M$. There is clearly a map $\overline{\alpha} : E\left(\zeta^n\left(\mathbb{R}^N\right)\right) \to E\left(\zeta^n\left(\mathbb{R}^M\right)\right)$ such that $(\overline{\alpha}, \alpha)$ is a bundle map from $\zeta^n\left(\mathbb{R}^N\right)$ to $\zeta^n\left(\mathbb{R}^M\right)$ and thus $\zeta^n\left(\mathbb{R}^N\right) \approx \alpha^*\left(\zeta^n\left(\mathbb{R}^M\right)\right)$.

In algebraic topology, one often considers the union $G_0(\mathbb{R}^\infty)$ of the increasing sequence $G_n\left(\mathbb{R}^{n+1}\right) \subset G_n\left(\mathbb{R}^{n+1}\right) \subset \cdots$ with weak topology; that is, a set $U \in G_n(\mathbb{R}^\infty) = \cup_l G_n\left(\mathbb{R}^{n+l}\right)$ is open if and only if $U \cap G_n\left(\mathbb{R}^{n+l}\right)$ is open in $G_n\left(\mathbb{R}^{n+l}\right)$ for all l. There is a natural n-dimensional bundle ζ^n over $G_n(\mathbb{R}^\infty)$ defined in a way similar to $\zeta^n\left(\mathbb{R}^N\right)$ such that the following properties are maintained: (i) for every bundle ξ over a paracompact space X, there is a map $f : X \to G_n(\mathbb{R}^\infty)$ such that $\xi \simeq f^*(\zeta^n)$. (ii) if $f_0, f_1 : X \to G_n(\mathbb{R}^\infty)$ are maps of a paracompact space X into $G_n(\mathbb{R}^\infty)$ with $f_0^* :$ $\zeta^n \simeq f_1^* \zeta^n$ then $f_1 \simeq f_0$.

For this reason ζ^n is called the *universal n-dimensional bundle* and $G_n(\mathbb{R}^\infty)$, is called the classifying space for n-dimensional bundles since equivalence classes of n-dimensional bundles over X are classified by homotopy classes of maps of X into $G_n(\mathbb{R}^\infty)$. Now $G_n(\mathbb{R}^\infty)$ is not a manifold so we continue to use the bundles $\zeta^n\left(\mathbb{R}^N\right)$, which are usually called universal bundles.

An orientation for a vector space V is an equivalence class of ordered bases for V where $(v_1, \ldots, v_n) \sim (w_1, \ldots, w_n)$ if and only if $\left(a_{ij}\right)$ defined by $w_i = \sum_j a_{ji} v_j$ has $\det\left(a_{ij}\right) > 0$. There are only two such equivalence classes η and $-\eta$. An oriented vector space is a pair (V, η), where η is an orientation for V.

An orientation for a bundle $\xi = \pi : E \to X$ is a collection $\eta = \{\eta_x\}$ of orientations for the fibres $\pi^{-1}(x)$ which satisfy an obvious compatibility requirement, while an oriented bundle is a pair (ξ, η), where η is an orientation for ξ. Orientation η of ξ gives another $-\eta = \{-\eta\}$ if X is connected. This is the only other one for ξ. Define $(\xi_1 \oplus \xi_2, \mu_1 \oplus \mu_2)$ to be the sum $\xi_1 \oplus \xi_2$ with the indicated orientation.

Suppose $\xi = \pi : E \to M$ is a smooth oriented n-dimensional vector bundle over a smooth manifold M of any dimension. The Euler class $\chi(\xi) \in H^n(M)$ was defined by first defining the Thom class $U(\xi) \in H_c^n(E)$. It can be proved $U(\xi)$ is the unique class whose restriction to each $\pi^{-1}(p)$ is the generator $\nu_p = H_c^n(\pi^{-1}(p))$ determined by the orientation. This result leads directly into the next theorem.

Theorem 4.1. Let $\xi = \pi : E \to M$ be a smooth manifold where M' is also a compact manifold. If E is the total space of $f^* \xi$ and $\tilde{f} : E' \to E$ is a bundle map covering f,

$$\tilde{f}^* (U(\xi)) = U(f^* \xi) \in H_c^n(E'). \tag{28}$$

Proof: Note $\tilde{f}$ has the property inverse of a compact set is compact, so $\tilde{f}^*$ takes $H_c^n(E)$ to $H_c^n(E')$. Let $f^* \xi$ be $\pi' : E' \to M'$. If $p \in M'$ is any point, and $i_{p'} : \pi'^{-1}(p) \to E'$ is the inclusion map, then

$$i_{p'}^* \tilde{f} U(\xi) = \left(\tilde{f} \circ i_{p'}\right)^* U(\xi). \tag{29}$$

Recall how $f^* \xi$ is defined then $\left(\tilde{f} \circ i_{p'}\right)^* U(\xi)$ must be the generator of $H_c^n(\pi'^{-1}(p'))$, since $i_{f(p')}^* U(\xi)$ is the generator of $H_c^n(\pi^{-1}(f(p')))$. This shows $\tilde{f}^* U(\xi)$ must be $U(f^* \xi)$. □

The Euler class $\chi(\xi)$ was defined as $s^* U(\xi)$ for any section s of ξ. Suppose $s = 0$ is the zero section, which is chosen. It can be shown that

$$f^* \chi(\xi) = \chi(f^* \xi) \in H^n(M'). \tag{30}$$

A consequence of (30) is important as it gives the following.

Theorem 4.2. If n is even, then

$$\chi\left(\tilde{Q}^n(\mathbb{R}^N)\right) \neq 0, \qquad N > n. \tag{31}$$

Proof: Since $S^n \subset \mathbb{R}^N$ for $N > n$, we have a bundle map $\left(\tilde{f}, f\right) : TS^n \to E\left(\tilde{\zeta}^n(\mathbb{R}^N)\right)$ so

$$\chi(TS^n) = f^* \chi\left(\tilde{\zeta}^n(\mathbb{R}^N)\right). \tag{32}$$

However, it is known to be the case that $\chi(TS^n)$ is $\chi(S^n)$ times the fundamental class of S^n and $\chi(S^n) = 2 \neq 0$. □

5. A unique one-form constructed from the curvature

This is an important characteristic class which is important and plays a significant role. Consider principal bundles associated with a smooth oriented n-dimensional vector bundle ξ over a smooth manifold M. There is the principal bundle of frames $F(\xi)$ of E. If ξ has a Riemannian metric $\langle , \rangle$, the bundle $O(E)$ of orthonormal frames can be considered, which is a principal bundle with group $O(n)$. Since only paracompact M are considered here, there is an Ehresmann connection ω on $O(E)$. Thus ω is a matrix of one-forms $\left(\omega_j^i\right)$ on $O(E)$ with values in $o(n)$, the curvature form $\Omega = D\omega$ is a matrix of two-forms $\left(\Omega_j^i\right)$ with values in $o(n)$. A connection ω on $O(E)$ is equivalent to a covariant derivative operator on E compatible with the metric, and for a general ξ over M, there will be many connections compatible with the metric. One can not be singled out by requiring a symmetric connection which only makes sense for the tangent bundle. As ξ is oriented, we can also consider the bundle $SO(E)$ of positively oriented frames. If X is connected, it is simply one of the two components of $O(E)$, with group $SO(n)$ and Lie algebra $o(n)$. A connection ω on $SO(n)$ again has values in $o(n)$ as does the matrix of two-forms Ω.

If we specialize to the case of a smooth oriented n-dimensional vector bundle $\xi = \pi : E \to M$ over M, with $n = 2m$ even. If $\langle , \rangle$ is a Riemannian metric for ξ and ω is a connection on the corresponding principal bundle $\overline{\omega} : SO(E) \to M$, consider the n-form which is defined on $SO(E)$

$$2m \cdot m!\ \mathrm{Pf}(\Omega) = \sum_{i_1, \cdots, i_n} \varepsilon^{i_1, \cdots, i_n} \Omega_{i_2}^{i_1} \wedge \cdots \wedge \Omega_{i_n}^{i_{n-1}}. \tag{33}$$

The following is an invariant formulation of a previous theorem.

Theorem 5.1. There is a unique n-form Λ on M such that

$$\overline{\omega}^* (\Lambda) = \sum_{i_1, \cdots, i_n} \varepsilon^{i_1, \cdots, i_n} \Omega_{i_2}^{i_1} \wedge \cdots \wedge \Omega_{i_n}^{i_{n-1}} = 2^m m!\ \mathrm{Pf}(\Omega). \tag{34}$$

Proof: Let $X_1, \cdots, X_n \in M_p$ $Y_1, \cdots, Y_n \in SO(E)_u$ be tangent vectors such that $\pi Y_i \to X_i$, and choose some $u \in \omega^{-1}(p)$. Then form Λ must satisfy

$$\Lambda(X_1, \cdots, X_n) = 2^m \cdot m!\ \mathrm{Pf}(\Omega)(Y_1, \cdots, Y_n) \tag{35}$$

This suffices to give uniqueness. If it can be shown this Λ in (35) is well-defined, then existence can be established.

Suppose different tangent vectors $Z_1, \cdots, Z_n$ are taken such that $\overline{\omega}_x Z_i = X_i$. Since $\overline{\omega}_x (Y_i - Z_i) = 0$, all $Y_i - Z_i$ are vertical. However, $\Omega(Y, Z) = 0$ if either Y or Z is vertical. Consequently,

$$\begin{aligned}\mathrm{Pf}(Y_1, \cdots, Y_n) = \mathrm{Pf}(\Omega)(Z_1, Y_2, \cdots, Y_n) &= \mathrm{Pf}(\Omega)(Z_1, Z_2, Y_3, \cdots, Y_n) \\ &= \mathrm{Pf}(\Omega)(Z_1, \cdots, Z_n).\end{aligned} \tag{36}$$

This means the definition of Λ does not depend on the Y_i selected. Suppose a different $\overline{u} \in \overline{\omega}^{-1}(p)$ is chosen. Then $\overline{u} = R_A(u) = u \cdot A$ for some $A \in SO(n)$, and so let $\overline{Y}_i \in SO(E)_\pi$ be given by $\overline{Y}_i = R_{A*} Y_i$ and

$$\begin{aligned}\mathrm{Pf}(\Omega)\left(\overline{Y}_1, \cdots, \overline{Y}_n\right) = \mathrm{Pf}(\Omega)(R_{A*} Y_1, \cdots, R_{A*} Y_n) &= \mathrm{Pf}\left(R_A^* \Omega\right)(Y_1, \cdots, Y_n) \\ = \mathrm{Pf}\left(A^{-1} \Omega A\right)(Y_1, \cdots, Y_n) &= \mathrm{Pf}(\Omega)(Y_1, \cdots, Y_n).\end{aligned} \tag{37}$$

□

Theorem 5.2. The unique n-form Λ in (35) is closed, $d\Lambda = 0$.

Proof: Suppose $X_1, \cdots, X_{n+1} \in M_p$ be given and choose $u \in \overline{\omega}^{-1} p)$ and $Y_1, \cdots, Y_{n+1} \in SO(E)_u$ with $\overline{\omega}_x Y_i = X_i$ and $h Y_i$ the horizontal component of Y_i. Then working out $d\Lambda$

$$d\Lambda(X_1, \cdots, X_{n+1}) = d\Lambda(\overline{\omega}_x Y_1, \cdots, \overline{\omega}_x Y_{n+1}) = d\Lambda(\overline{\omega}_x hY_1, \cdots, \overline{\omega}_x hY_{n+1})$$
$$= (\overline{\omega}^* d\Lambda)(hY_1, \cdots, hY_n) = d(\overline{\omega}^* \Lambda)(hY_1, \cdots, hY_{n+1}) \tag{38}$$
$$= 2^m \cdot m! d(\mathrm{Pf}(\Omega))(hY_1, \cdots, hY_{n+1}) = 2^m \cdot m! D(\mathrm{Pf}(\Omega))(Y_1, \cdots, Y_{n+1}).$$

However, $D\Omega = 0$ by Bianchi's identity and a consequence of this is that (38) vanishes. □

This result applies automatically when ξ is the tangent bundle. The implication of this is that the n-form Λ determines a de Rham cohomolgy class $[\Lambda] \in H^n(M)$ of M. The form Λ itself depends on the oriented n-dimensional bundle $\xi = \pi : E \to M$ over M as well as the choice of metric for ξ and connection ω on the corresponding bundle $SO(E)$.

Theorem 5.3. The cohomology class $[\Lambda]$ is independent of both the metric and the connection ω.

Proof: Suppose two metrics $\langle , \rangle$, $\langle , \rangle'$ are given for ξ. Then the corresponding principal bundles $SO(E)$ and $SO'(E)$ are equivalent. If $\tilde{f} : SO'(E) \to SO(E)$ is a fiber preserving diffeomorphism which commutes with the action $SO(n)$ and ω a connection on $SO(E)$. Then $\omega' = \tilde{f}^* \omega$ is a connection on $SO(E)$. Corresponding curvature forms satisfy $\Omega' = \tilde{f}^* \, \mathrm{Pf}(\Omega)$ so $\mathrm{Pf}(\Omega') = \tilde{f}^* \, \mathrm{Pf}(\Omega)$. The corresponding forms Λ and Λ' are in fact equal. It suffices to show any two connection differential forms ω_0, ω_1 on the same $SO(E)$ generate forms Λ_0, Λ_1 whose difference is exact. If $\pi : M \times [0,1] \to M$ is the projection $\pi(p,t) = p$, consider the bundle $\pi^* SO(\xi)$ over $M \times [0,1]$. Induced connections are $\pi^* \omega_0$ and $\pi^* \omega_1$ on $\pi^* SO(\xi)$. Let $\tau : M \times [0,1] \to [0,1]$ defined here as $\tau(p,t) = t$ and define a connection

$$\omega = (1 - \tau)\pi^* \omega_0 + \tau \pi^* \omega_1 \tag{39}$$

on $\pi^* SO(\xi)$ with Ω the connection form. If i_t maps M to $M \times [0,1]$ and is defined as $i_t(p) = (p,t)$, then $i_0^*(\omega)$ can be identified with ω_0 and $i_1^*(\omega)$ with ω_1. By Theorems 5.1 and 5.2, which hold for manifolds with and without boundary, there is a closed n-form Λ on $M \times [0,1]$ which pulls back to $2^m m! \, \mathrm{Pf}(\Omega)$ on the total space of $\pi^* SO(\xi)$. A theorem states for any k-form ω on $M \times [0,1]$, $i_1^* \omega - i_0^* \omega = d(I\omega) - I(d\omega)$. So if $d\omega = 0$, this implies $i_1^* \omega - i_0^* \omega = d(I\omega)$. Substituting the form Λ in place of ω into this, it follows that $\Lambda_1 - \Lambda_0$ is exact. □

Thus every smooth oriented smaooth bundle ξ over M of even fibre dimension n determines a de Rham cohomology class $C(\xi) = [\Lambda] \in H^n(M)$ and $C(\xi) = C(\eta)$ if $\xi \simeq \eta$. It may be asked how does the object $C(\xi)$ behave with respect to f^*.

Theorem 5.4. Let $\xi = \pi : \xi \to M$ be a smooth oriented bundle over M with fibre dimension n even, let $f : M' \to M$ be a smooth map. Then

$$C(f^* \xi) = f^* C(\xi) \in H^n(M'). \tag{40}$$

Proof: The total space of $f^* \xi$ is called E'. Let $\tilde{f} : E' \to E$ be the bundle map covering f. If $\langle , \rangle$ is a metric on E, then $\tilde{f}^* \langle , \rangle$ is a metric on E. There is an equivalence $\overline{f}$: $SO(E') \to SO(E)$ covering f with $\overline{\omega}'$ taking $SO(E')$ to M' and $\overline{\omega}$ mapping $SO(E)$ to M.

If ω is a connection on $SO(E)$, then $\overline{f}^*(\omega)$ is a connection on $SO(E')$. It is seen that the corresponding connection forms satisfy $\Omega' = \overline{f}^* \Omega$. Aa a result, we have

$$\mathrm{Pf}(\Omega') = \mathrm{Pf}\left(\overline{f}^* \Omega\right) = \overline{f}^* \, \mathrm{Pf}(\Omega). \tag{41}$$

For n-forms Λ on M given by Theorem 5.1, we then have

$$\overline{\omega}'^*(f^*\Lambda) = \overline{f}^*\overline{\omega}^*\Lambda = 2^m \cdot m!\overline{f}^* \operatorname{Pf}(\Omega) = 2^m \cdot m! \operatorname{Pf}(\Omega'). \tag{42}$$

This means $f^*\Lambda$ must be the n-form Λ' on M' given in (31). □

When ξ is a smooth oriented bundle of odd fibre dimension, the definition of C may be extended. It would be remarkable if it were the case that $C(\xi)$ were always a constant multiple of $\chi(\xi)$. To this end, the following theorem is needed.

Theorem 5.5. Let $\xi_i = \pi_i : E_i \to M$ for $i = 1, 2$ be smooth oriented vector bundles over M of fibre dimension n_1 and n_2. If $n_i = 2m_i$, then

$$C(\xi_1 \oplus \xi_2) = \frac{(m_1 + m_2)!}{m_1! m_2!} C(\xi_1) \cup C(\xi_2). \tag{43}$$

If n_1 or n_2 is odd, this reduces to $C(\xi_1 \oplus \xi_2) = 0$.

Proof: Pick two metrics which are Riemannian for each ξ_i and set $\langle \cdot, \cdot \rangle = \langle \cdot, \cdot \rangle_1 \oplus \langle \cdot, \cdot \rangle_2$ on $\xi_1 \oplus \xi_2 = \pi : E \to M$. Let $\overline{\omega}_i : SO(E_i) \to M$ and $\overline{\omega} : SO(E) \to M$ be the corresponding principal bundles. Over M consider the product principal bundle $Q = SO(E_1) * SO(E_2)$ with corresponding group $SO(n_1) \times SO(n_2) \subset SO(n_1 + n_2)$ whose fiber over $p \in M$ is the direct product $\overline{\omega}_1^{-1} \times \overline{\omega}_2^{-1}(p)$, so this bundle is a subset of $SO(E)$.

Let ρ_i be the projection maps for Q which project this down onto either of its factors. If ω_i are connections on $SO(E_i)$, with curvature forms Ω_i, then

$$\rho_1^* \omega_1 \oplus \rho_2^* \omega_2 = \begin{pmatrix} \rho_1^* \omega_1 & 0 \\ 0 & \rho_2^* \omega_2 \end{pmatrix} \tag{44}$$

is a connection on Q and the curvature form is

$$\Omega = \rho_1^* \Omega_1 \oplus \rho_2^* \Omega_2 = \begin{pmatrix} \rho_1^* \Omega_1 & 0 \\ 0 & \rho_2^* \Omega_2 \end{pmatrix} \tag{45}$$

The connection $\overline{\omega}$ can be extended uniquely to a connection $\tilde{\omega}$ on $SO(E)$. The requirement $\tilde{\omega}(\sigma(M)) = M$ determines $\tilde{\omega}$ at the new vertical vectors, hence $\tilde{\omega}$ is determined at all points of Q, and then at all points of $SO(E)$ by the requirement $\tilde{\omega}(R_A^* Y) = Ad(A^{-1})\tilde{\omega}(Y)$.

At any point $e \in Q$, the horizontal vectors for $\tilde{\omega}$ are the same as that for $\overline{\omega}$. At E, it holds that $\overline{\Omega} = \tilde{\Omega}$ for tangent vectors to Q which implies, using $\operatorname{Pf}(A \oplus B) = \operatorname{Pf}(A) \cdot \operatorname{Pf}(B)$, that

$$\operatorname{Pf}(\tilde{\Omega}) = \operatorname{Pf}(\overline{\Omega}) = \operatorname{Pf}(\rho_1^* \Omega_1) \wedge \operatorname{Pf}(\rho_2^* \Omega_2) = \rho_1^* \operatorname{Pf}(\Omega_1) \wedge \rho_2^* (\Omega_2). \tag{46}$$

Consequently, if Λ_i are the forms given by (34), then at e it must hold that on tangent vectors to Q

$$\begin{aligned} \overline{\omega}^* \Lambda = 2^{m_1 + m_2} (m_1 + m_2)! \operatorname{Pf}(\tilde{\Omega}) &= \frac{(m_1 + m_2)!}{m_1! m_2!} 2^{m_1} m_1! \rho_1^* \operatorname{Pf}(\Omega_1) \wedge 2^{m_2} m_2! \rho_2^* \operatorname{Pf}(\Omega_2) \\ &= \frac{(m_1 + m_2)!}{m_1! m_2!} \rho_1^* \overline{\omega}_1^* \Lambda_1 \wedge \rho_2^* \overline{\omega}_2^* \Lambda_2 = \frac{(m_1 + m_2)!}{m_1! m_2!} \overline{\omega}_1^* \Lambda_1 \wedge \overline{\omega}_2^* \Lambda_2. \end{aligned} \tag{47}$$

This implies that

$$\Lambda = \frac{(m_1 + m_2)!}{m_1! m_2!} \Lambda_1 \wedge \Lambda_2. \tag{48}$$

□

Corollary 5.1. If the oriented bundle $\xi = \pi : E \rightarrow M$ has a nowhere zero section s, then

$$C(\xi) = 0. \tag{49}$$

Proof: Let $E_1 \subset E$ be written

$$\cup_{p \in M} \mathbb{R} \cdot s(p), \tag{50}$$

and let $E_2 \subset E$ be the orthogonal complement

$$\cup_{p \in M} (\mathbb{R} \cdot s(p))^{\perp} \tag{51}$$

with respect to some Riemannian metric on E. Then $\xi_1 = \pi_1 | E_1 : E_1 \rightarrow M$ is an oriented one-dimensional bundle. Consequently, $\xi_2 = \pi_2 | E_2 : E_2 \rightarrow M$ is also an oriented bundle since ξ is oriented. Clearly $\xi \simeq \xi_1 \oplus \xi_2$. An application of the previous result shows that $C(\xi) = 0$. □

This theorem is almost enough to characterize χ as we can now show the statement which relates $C(\xi)$ and the Euler class.

Corollary 5.2. If $\xi = \pi : E \rightarrow M$ is a smooth vector bundle of fibre dimension n over a compact oriented manifold M, then the class $C(\xi) \in H^n(M)$ is a multiple of the Euler class $\chi(\xi)$.

Proof: Suppose S is the sphere bundle $S = \{e \in E : \langle e, e \rangle = 1\}$, which is constructed with respect to some Riemannian metric on E. Let $\pi_0 : S \rightarrow X$ be the restriction $\pi | S$. The bundle $\pi_0^* \xi$ has a nowhere zero section. Corrollary 5.1 and Theorem 5.4 then yield

$$\pi_0^* C(\xi) = C(\pi_0^* \xi) = 0. \tag{52}$$

However, there is a theorem which states a class $\alpha \in H^n(M)$ satisfies $\pi_0^* \alpha = 0$ if and only if α is a multiple of $\chi(\xi)$. It can now be inferred that $C(\xi)$ is a multiple of the Euler class $\chi(\xi)$. □

6. The Gauss-Bonnet-Chern theorem

If Corollary 5.2 is applied to the tangent bundle of a compact oriented manifold M of dimension n which is even, the class $C(TM) \in H^n(M)$ is some multiple of the Euler class $\chi(TM)$. This fact is not so interesting because $H^n(M)$ is one-dimensional since it means $C(TM) = 0$ if $\chi(TM) = 0$. The corollary does lead to something interesting when applied to the universal bundle.

Theorem 6.1 For every even n, there is a constant β_n such that

$$C(\xi) = \beta_n \chi(\xi). \tag{53}$$

for all smooth oriented n-dimensional bundles ξ over compact oriented manifolds. In this sense, it is universal.

Proof: Begin with the bundles $\tilde{zeta}_n(\mathbb{R}^N)$ for $N > n$. Corollary 5.2 implies there are constants $\beta_{n,N}$ such that

$$C\left(\tilde{\zeta}^n(\mathbb{R}^N)\right) = \beta_{n,N}\chi\left(\tilde{\zeta}^n(\mathbb{R}^N)\right) \in H^n(\tilde{\zeta}_n(\mathbb{R}^N)). \tag{54}$$

If $j : \tilde{G}_n(\mathbb{R}^N) \to \tilde{G}_n(\mathbb{R}^M)$ is the natural inclusion, then $j^*\left(\tilde{\zeta}^N(\mathbb{R}^N)\right) \simeq \tilde{\zeta}^n(\mathbb{R}^N)$. Equation (30) and Theorem 5.4 yield

$$C\left(\tilde{\zeta}^n(\mathbb{R}^N)\right) = j^*C\left(\tilde{\zeta}^n(\mathbb{R}^M)\right) = j^*\chi\left(\tilde{\zeta}^n(\mathbb{R}^M)\right). \tag{55}$$

Thus, (54), (55) give

$$\beta_{n,N}\chi\left(\tilde{\zeta}^n(\mathbb{R}^N)\right) = \beta_{n,M}\chi\left(\tilde{\zeta}^n(\mathbb{R}^N)\right). \tag{56}$$

Since $\chi\left(\tilde{\zeta}^n(\mathbb{R}^N)\right) \neq 0$ by Theorem 4.2, this implies that $\beta_{n,N} = \beta_{n,M}$ for all $M, N > 1$. This common number is called β_n, and we have

$$C\left(\tilde{\zeta}^n(\mathbb{R}^N)\right) = \beta_n\chi\left(\tilde{\zeta}^n(\mathbb{R}^N)\right). \tag{57}$$

However it is known that any smooth oriented n-dimensional bundle ξ over a compact manifold M is equivalent to $f^*\left(\tilde{\zeta}^n(\mathbb{R}^N)\right)$ for some smooth map $f : M \to \tilde{G}(\mathbb{R}^N)$, then

$$C(\xi) = C\left(f^*\tilde{\zeta}^n(\mathbb{R}^N)\right) = f^*C\left(\tilde{\zeta}^n(\mathbb{R}^N)\right) = \beta_n f^*\chi\left(\tilde{\zeta}^n(\mathbb{R}^N)\right) = \beta_n\chi(\xi). \tag{58}$$

□

The constant β_N is universal in nature and it may be asked whether it can be computed. Since it has this universality property, it suffices to compute this constant for a special case where the calculation is easier and in turn implies another application of the next theorem.

Theorem 6.2 For integer $n = 2m$, the constant β_n in Theorem 6.1 is

$$\beta_n = \frac{n!}{2}V(S^n) = \pi^m 2^n m!. \tag{59}$$

If $(M, \langle,\rangle)$ is a compact manifold of even dimension $n = 2m$ then

$$\int_M K_n\, d\mu_g = \frac{\pi^m 2^n m!}{n!}\chi(M). \tag{60}$$

Proof: Let ξ be the tangent bundle TM of a compact oriented manifold of dimension n. Now (17) gives a formula for $K_n\, d\mu_g$, where Ω^i_j are curvature forms for some positively oriented orthonormal moving frame. This implies that the form Λ in Theorem 5.1 for the bundle $SO(\xi) = SO(TM)$ is

$$\Lambda = n! K_n \, d\mu_g. \tag{61}$$

If κ is the fundamental class of M then

$$\left(\int_M K_n \, d\mu_g \right) \kappa = \frac{1}{n!} \left(\int_M \Lambda \right) \kappa = \frac{1}{n!} C(\xi) = \frac{\beta_n}{n!} \chi(\xi) = \frac{\beta_n}{n!} \chi(M) \cdot \kappa. \tag{62}$$

Hence, equating the coefficients of κ on both sides,

$$\int_M K_n \, d\mu_g = \frac{\beta_n}{n!} \chi(M). \tag{63}$$

Consider the case of a specific manifold $M = S^n$ in (63), where $K_n = 1$ so the left side of (63) reduces to $V(S^n)$

$$V(S^n) = \frac{\beta_n}{n!} \chi(S^n) = \frac{2\beta_n}{n!}. \tag{64}$$

Since the volume $V(S^n)$ is known to be $\pi^m 2^{n+1} m!/n!$, solve (64) for β_n in terms of $V(S^n)$,

$$\beta_n = \frac{n!}{2} V(S^n) = \pi^m 2^n m!. \tag{65}$$

This value of β_n can be put back into (60) and for the manifold M, it follows that

$$\int_M K_n \, d\mu_g = \frac{\pi^m 2^n m!}{n!} \chi(M). \tag{66}$$

□

7. The theorem for manifolds with boundary

Theorem 5.5 played a large part in the proof of (64). It allowed us to state that if $\xi = \pi : E \to M$ is an oriented n-dimensional vector bundle with sphere bundle $\pi_0 : S_0 \to M$, then $\pi_0^* C(\xi) = 0$ was a large part of the proof of (). If Λ is the n-form on M representing $C(\xi)$, then the n-form $\pi_0^* \Lambda$ on S is exact $\pi_0^* \Lambda = d\Phi$ for some $(n-1)$-form on S. Suppose $\xi = TM^*$ and let X be a unit vector bundle on M which has an isolated singularity at $p \in M$ Let $B(\varepsilon)$ be a closed ball of radius ε around p and set $M_\varepsilon = M \backslash \text{int}(B(\varepsilon))$ where int denotes the interior. Then $X(M_\varepsilon)$ is a manifold with boundary, the image of M_ε under the section $X : M \backslash \{p\} \to S$. Consequently

$$\int_M \Lambda = \int_{M-\{p\}} \Lambda = \lim_{\varepsilon \to 0} \int_{M_\varepsilon} \Lambda = \lim_{\varepsilon \to 0} \int_{M_\varepsilon} X^* \left(\pi_0^* \Lambda \right) = \lim_{\varepsilon \to 0} \int_{X(M_\varepsilon} \pi_0^* \Lambda$$

$$= \lim_{\varepsilon \to 0} \int_{X(M_\varepsilon)} d\Phi = \lim_{\varepsilon \to 0} \int_{\partial X(M_\varepsilon)} \Phi. \tag{67}$$

If $\text{ind}(X,p)$ is the index of X at p

$$\int_M \Lambda = \text{ind}(X,p)\int_{\pi_0^{-1}(p)} \Phi = \chi(M)\int_{\pi_0^{-1}(p)} \Phi. \tag{68}$$

Since $n = 2m$ we also have the Gauss-Bonnet-Chern Theorem 6.2

$$\int_M \Lambda = \int_M n! K_n \, d\mu_g = \pi^m m! 2^n \chi(M), \tag{69}$$

we finally obtain

$$\int_{\pi_0^{-1}(p)} \Phi = \pi^m m! 2^n. \tag{70}$$

Let $(M, \partial M)$ be a compact orientable manifold with boundary with Euler characteristic $\chi(M) = \dim H^0(M) - \dim H^1(M) + \cdots$. A compact oriented manifold M^2 can be constructed, the double of M, by taking two disjoint copies of M and identifying corresponding points of ∂M.

Theorem 7.1 The Euler characteristic of the manifold M^2 is given by

$$\chi(M^2) = 2\chi(M) - \chi(\partial M). \tag{71}$$

Proof: Let U and V be open neighborhoods of the two copies of M in M^2 such that $H^k(U) \equiv H(V) \equiv H^k(M)$ for all k and $H^k(U \cap V) \equiv H^k(\partial M)$ for all k. So there is the sequence $0 \to H^0(M^2) \to \cdots \to H^k(M^2) \to H^k(U) \oplus H^k(V) \to \cdots \to H^k(U \cap V) \to H^{k+1}(M^2)$. When the sequence is exact, a theorem can be applied to obtain the result. □

This is very interesting since it claims different things depending on whether the dimension n of M is even or odd. When n is odd $\chi(M^2) = 0$ hence $\chi(M) = 1/2\chi(\partial M)$ which implies $\chi(\partial M)$ must be even. But when n is even, $\chi(\partial M) = 0$, so the previous theorem implies

$$2\chi(M) = \chi(M^2). \tag{72}$$

Corollary 7.1 Let M be a compact orientable manifold with boundary of even dimension n. Let X be a vector field on M with only finitely many zeros all in $M \setminus \partial M$ such that X is outward pointing on ∂M. The sum of indices of X is $\chi(M)$.

Proof: Modify X near ∂M so it is an outward pointing unit normal ν on the boundary and so there are no new zeros. Then there is a vector field on M^2 which looks like X on one copy of M and $-X$ on the other. Since n is even, the index $-X$ of an isolated zero is the same as the index of X at that zero. The Theorem of Poincaré-Hopf on the sum of indices of X gives twice the sum of the indices of X equals $\chi(M^2) = 2\chi(M)$ by (72). □

Theorem 7.2 Let M be a compact oriented Riemannian manifold with boundary of even dimension $n = 2m$, tangent bundle $\pi : TM \to M$ and associated sphere bundle $\pi_0 = \pi|S : S \to M$. Let ω be a connection on the principal bundle $\varpi : SO(TM) \to M$, with curvature form Ω. Let Λ be the unique n-form on M with

$$\varpi^* \Lambda = \sum \varepsilon^{i_1, \cdots, i_n} \Omega_{i_2}^{i_1} \wedge \cdots \wedge \Omega_{i_n}^{i_{n-1}} = 2^m m! \text{ Pf}(\Omega), \tag{73}$$

and Φ an $(n-1)$-form on S with $\pi_0^* \Lambda = d\Phi$. Let $\nu : \partial M \to S$ be the outward pointing unit normal on ∂M. Then

$$\int_M K_n \, d\mu_g = \frac{1}{n!} \int_M \Lambda = \frac{\pi^m m! 2^n}{n!} \chi(M) + \frac{1}{n!} \int_{\partial M} \nu^* \Phi. \tag{74}$$

Proof: Extend ν to a vector field X on M with only finitely many zeros $p_1, \ldots, p_k \in M \backslash \partial M$. Let $B_i(\varepsilon)$ be the closed balls of radius ε centered at p_i which are disjoint from each other and from ∂M. Put $M_\varepsilon = M \backslash \cup_{i=1}^k \operatorname{int} B_i(\varepsilon)$. Now integrate the form Λ over M and use (70)

$$\begin{aligned} \int_M \Lambda &= \lim_{\varepsilon \to 0} \int_{\partial X(M_\varepsilon)} \Phi = \int_{\nu(\partial M)} \Phi + \sum_{i=1}^k \lim_{\varepsilon \to 0} \int_{\partial B_i(\varepsilon)} \Phi \\ &= \int_{\partial M} \nu^* \Phi + \pi^m m! 2^n \sum_{i=1}^M (\operatorname{ind} X)_{p_i} = \int_{\partial M} \nu^* \Phi + \pi^m m! 2^n \chi(M). \end{aligned} \tag{75}$$

The last line makes use of Corollary 7.1. □

Theorem 7.2 presents one way in which Theorem 6.2 can be generalized to the case of manifolds with boundary. At this point an interpretation for the first term in (75) is not easy to provide. It is required to obtain an explicit Φ such that $\pi_0^* \Lambda = d\Phi$. In fact such a Φ can be constructed.

Author details

Paul Bracken
Department of Mathematics, University of Texas, Edinburg, TX, USA

*Address all correspondence to: paul.bracken@utrgv.edu

References

[1] Chern SS. A simple intrinsic proof of the Gauss-Bonnet formula for closed Riemannian manifolds. Annals of Mathematics. 1944;**45**(2):747-752

[2] Chern SS. On curvature and characteristic classes of a Riemannian manifold. Abhandlungen aus dem Mathematischen Seminar der Universität Hamburg. 1955;**20**:117-126

[3] Bishop RL, Goldberg SI. Some implications of the generalized gauss-bonnet theorem. Transactions of the American Mathematical Society. 1964; **112**(3):508-535

[4] Bell D. The Gauss-Bonnet Theorem for vector bundles. Journal of Geometry. 2006;**85**:15-21

[5] Peterson P. Riemannian Geometry. New York: Springer-Verlag; 1998

[6] Goldberg SI. Curvature and Homology. New York: Academic Press; 1962

[7] Spivak M. A Comprehensive Introduction to Differential Geometry. Houston: Publish or Perish; 1999

[8] Gilkey P. Invariance Theory, the Heat Equation, and the Atiyah-Singer Index Theorem. 2nd ed. Boca Raton: CRC Press; 1995

[9] Boothby W. An Introduction to Differentiable Manifolds and Riemannian Geometry. New York: Academic Press; 1975

[10] Kobayashi S, Nomizu K. Foundations of Differential Geometry. New York: Interscience; 1963

[11] Rosenberg S. The Laplacian on a Riemannian Manifold. Cambridge: Cambridge University Press; 1997

[12] Glavan D, Lin C. Einstein-Gauss-Bonnet gravity in four-dimensional spacetime. Physical Review Letters. 2020;**124**:081301

[13] Herrero-Valea M. The shape of scalar Gauss-Bonnet gravity. Journal of High Energy Physics. 2022;**13**(03):075

[14] Liu Y, Pan Q, Wang B, Cai R-G. Dynamical perturbations and critical phenomena in Gauss-Bonnet AdS black holes. Physics Letters A. 2010;**693**: 343-350

[15] D'Hoker E, Phong DH. The geometry of string perturbation theory. Reviews of Modern Physics. 1988;**60**(**4**): 917-1065

[16] Chow B, Lu L, Ni L. Hamilton's Ricci Flow. Providence, RI: AMS; 2006

Section 2

Applications of Manifolds to Physics

Chapter 4

Smooth Structures on Spin Manifolds in Four Dimensions

Simon Davis

Abstract

The estimate of the coefficient of the magnitude of the signature, which are defined by the number of positive and negative eigenvalues in the inequality representing smooth, oriented, simply connected, compact, spin four-manifolds with indefinite intersection forms can be increased until it is equal to the conjectured value. Therefore, if the intersection form is $mE_8 \oplus n\begin{pmatrix} 0 & 1 \\ 1 & 0 \end{pmatrix}$, the oriented, simply connected, compact, spin four-manifold will admit a smooth structure if and only if $n \geq \frac{3}{2}|m|$. The inequality is changed to $n \geq \frac{3}{2}|m| - \left(1 - \frac{1}{2^i}\right)$, there is a 2^i-fold spin covering of a non-spin manifold M given the demonstration of $n \geq \frac{3}{2}|m|$ for oriented, compact, spin manifolds. A closer examination of the proof reveals that the lower bound for b_+ can be increased to $3|k| + 1$, where $|k| = \frac{3}{15}|\sigma|$ for a spin manifold, yielding $b_2 \geq \frac{11}{8}|\sigma| + 2$. The projection of a spin covering to a non-spin manifold yields the lower bound $b_2 \geq \frac{11}{8}|\sigma|$, which establishes the prediction for the coefficients of intersection forms for this class of smooth, oriented, simply connected, compact four-manifolds.

Keywords: intersection forms, coefficients, spin manifolds, smooth structures

1. Introduction

The classification of four-manifolds may be determined by the handlebody decomposition into simply connected components of a topological sum when the manifold is smooth. If it is closed, oriented, and simply connected, then it will be distinguished, within a homotopy equivalence, by an intersection form that is either definite, indefinite with odd parity, or indefinite with even parity. These manifolds also be identified by the second Betti number and the signature. The four-manifold admits a smooth structure if the intersection matrix is definite or indefinite with odd parity. Furthermore, if the intersection form is indefinite and equals $mE_8 \oplus n\begin{pmatrix} 0 & 1 \\ 1 & 0 \end{pmatrix}$, it will continue to have a smooth structure if $n \geq \frac{3}{2}|m|$. In the (b_2, σ) plane, the smooth structures are located above the line $b_2 \geq \frac{11}{8}|\sigma|$ [1] and the nonsmooth structures are located below the line $b_2 \leq \frac{5}{4}|\sigma|$, with the region between the two lines undetermined. It may be shown, however, that one of the manifolds in this region does not admit a smooth structure [2]. The coefficient of $\frac{11}{8}$ also will be the maximal value for the line separating the set of manifolds with smooth and nonsmooth structures because the inequality is saturated by $K3$ [3]. Consequently, it remains to be established that all of the manifolds with an indefinite intersection form in this intermediate region do not admit a smooth structure. The condition of an indefinite intersection form is necessary because $b_2 = 1$ and $\sigma = 1$ for $\mathbb{CP}^2$.

By considering finite-dimensional approximations to the Seiberg-Witten map of the tensor product sections of the spinor bundle and the space of connections, the lower bound $b_2 \geq \frac{10}{8}|\sigma| + 2$ was established for oriented, connected spin manifolds [4]. It may be increased with the use of stable homotopy groups of spheres and Pin_2-equivariant homotopy invariants [5] to $b_2 \geq \frac{10}{8}\sigma + 4$. The lower limit will be increased first to $\frac{b_2}{|\sigma|} > \frac{21}{16}$ for spin manifolds with signature $\sigma \geq 16$ as a result of a theorem on the nonexistence of smooth four-manifolds with the intersection form $+4E - 8 \oplus 5H$ [6]. Then, the coefficient of $\frac{11}{8}$ will be found by considering precisely the form of the maps between finite-dimensional vector bundles over the four-manifold. A second proof will be derived by considering intersection products of second homology classes representable by spheres [7]. Consequences of the related $\frac{3}{2}$ conjecture for the embedding of surfaces with a nonvanishing second homology class in an irreducible four-manifold will be described.

2. The inequality for the second Betti number and the signature

Since the intersection matrix is symmetric and diagonalizable over $\mathbb{R}$, b_+ and b_- will denote the number of positive and negative eigenvalues respectively. Then the second Betti number and the signature are $b_2 = b_+ + b_-$ and $\sigma = b_+ - b_-$ respectively. Let $k = -\frac{\sigma(M)}{16}$ and the inequality $b_+ \geq 3k = -\frac{3\sigma}{16}$. When the signature is negative, and $-\sigma$ may be replaced by $|\sigma|$,

$$\begin{aligned} \frac{b_2 + \sigma}{2} &\geq \frac{3|\sigma|}{16} \\ b_2 &\geq 2\left[-\frac{\sigma}{2} + \frac{3|\sigma|}{16}\right] = \frac{11}{8}|\sigma|. \end{aligned} \tag{1}$$

The signature could be positive such that $b \geq 3\,k$ is a much less stringent inequality. However, by reversing the orientation, the sign of the signature is changed, and this inequality always can be taken to imply $b_+ \geq \frac{3|\sigma|}{16}$.

Similarly, if the orientation is chosen such that $b_+ \geq 2k + 1 = -\frac{\sigma}{8} + 1$ is equivalent to $b_+ \geq \frac{|\sigma|}{8} + 1$,

$$\begin{aligned} \frac{b_2 + \sigma}{2} &\geq \frac{|\sigma|}{8} + 1 \\ b_2 &\geq 2\left[\frac{|\sigma|}{2} + \frac{|\sigma|}{8} + 1\right] = \frac{5}{4}|\sigma| + 2. \end{aligned} \tag{2}$$

It may be demonstrated that a spin 4-manifold can admit a smooth structure when the intersection form is $4E_8 \oplus nH$ for $n \geq 6$ [7]. Consequently, there is no smooth manifold with the intersection form $4E_8 \oplus 5H$. Since the coefficients are relatively prime, the ratio $\frac{n}{m} = \frac{5}{4}$ is achieved only for the intersection forms $4kE_8 \oplus 5kH$, k = 1, 2, 3, ... Given that there is a manifold M with the intersection form $4E_8 \oplus 5H$, the latter form would characterize $\#kM$.

Lemma 2.1. Smooth, oriented, simply connected, compact 4-manifolds with a spin structure and an indefinite intersection form have second Betti numbers bounded by $\frac{21}{16}|\sigma|$, which is a bound closer to the line with gradient $\frac{11}{8}$ for non-zero signature.

Proof. The line representing smooth structures must be $n \geq \frac{5}{4}m$. Then

$$\frac{b_2}{|\sigma|} = \frac{8m + 2n}{8m} \geq \frac{8m + \frac{5}{2}m}{8m} = \frac{21}{6}. \tag{3}$$

When

$$\frac{10}{8} + \frac{2}{|\sigma|} \leq \frac{21}{16}, \tag{4}$$

or

$$|\sigma| \geq \frac{32}{5}, \tag{5}$$

this bound is better than the established value. If

$$\frac{10}{8} + \frac{4}{|\sigma|} < \frac{21}{16}, \tag{6}$$

or

$$|\sigma| \geq \frac{65}{5}. \tag{7}$$

Any spin manifold will have a signature divisible by 16 by Rohlin's theorem. Therefore, if it is non-zero, these inequalities will be valid.

The line with gradient $\frac{21}{16}$ in the geography of four-manifolds is closer to the boundary between smooth and nonsmooth structures.

Given the validity of the 11/8 conjecture, smooth connected spin four-manifolds can be regarded as the topological sums $\#kK3\#\ell S^2 \times S^2$ or $\#k\mathbb{CP}^2\#\ell\overline{\mathbb{CP}^2}$.

Lemma 2.2. All manifolds $\#kK3\#\ell S^2 \times S^2$ with $k > 0$ have $\frac{b_2}{|\sigma|} \geq \frac{11}{8}$, with the bound saturated by $K3$. The coefficients in $\#k\mathbb{CP}^2\#\ell\overline{\mathbb{CP}^2}$ must satisfy the inequalities $\frac{3}{19}k \leq \ell\,\frac{19}{3}k$ for a smooth structure to exist by the $\frac{11)}{8}$ conjecture.

Proof. Since $K3$ has an interesection form with 3 positive and 19 negative eigenvalues, $b_2(K3)$ = 22 and $\sigma(K3)$ = -16. The intersection form of $S^2 \times S^2$, H, has the eigenvalues 1 and $-$ 1, and $b_2(S^2 \times S^2)$ = 2, while $\sigma(S^2 \times S^2)$ = 0. Then

$$\begin{aligned} b_2\left(\#kK3\#\ell S^2 \times S^2\right) &= 22k + 2\ell \\ \sigma\left(\#kK3\#\ell S^2 \times S^2\right) &= -16k \end{aligned} \tag{8}$$

and

$$\frac{b_2(\#kK3\#\ell S^2 \times S^2)}{|\sigma(\#kK3\#S^2 \times S^2)|} = \frac{11}{8} + \frac{1}{8}\frac{\ell}{k}. \tag{9}$$

The intersection matrix of $\#k\mathbb{COP}^2\#\ell\overline{\mathbb{CP}^2}$ is $diag(1, ..., 1, -1, ..,-1)$ with

$$\begin{aligned} b_2\left(\#k\mathbb{CP}^2\#\ell\overline{\mathbb{CP}^2}\right) &= k + \ell \\ \sigma\left(\#k\mathbb{CP}^2\#\ell\overline{\mathbb{CP}^2}\right) &= k - \ell. \end{aligned} \tag{10}$$

It follows that

$$\frac{b_2\left(\#k\mathbb{CP}^2\#\ell\overline{\mathbb{CP}^2}\right)}{\sigma\left(\#k\mathbb{CP}^2\#\ell\overline{\mathbb{CP}^2}\right)} = \frac{k+\ell}{|k-\ell|} = \begin{cases} 1 + \dfrac{2\ell}{k-\ell} & k > \ell \\ 1 + \dfrac{2k}{\ell - k} & \ell > k \end{cases} \tag{11}$$

either

$$\begin{aligned} 1 + \frac{2\ell}{k-\ell} &\geq \frac{11}{8}, \\ \ell &\geq \frac{3}{19}k \end{aligned} \tag{12}$$

or

$$\begin{aligned} 1 + \frac{2k}{\ell-k} &\geq \frac{11}{8} \\ k &\geq \frac{3}{19}\ell \end{aligned} \tag{13}$$

which may combine in the inequalities

$$\frac{3}{19}k \leq \ell \leq \frac{19}{3}k. \tag{14}$$

A bound may be established for simply connected complex surfaces with an even cup product form [8]. It is known that, for these manifolds, $b_2 = c_2 - 2$ and $\sigma = \frac{1}{3}\left(c_1^2 - 2c_2\right)$, where c_1 and c_2 are the first two Chern numbers. Defining

$$b = \frac{1}{16}(8b_2 - 11|\sigma|), \tag{15}$$

the $\frac{11}{8}$ conjecture is equivalent to $b \geq 0$.When $\sigma < 0$,

$$\begin{aligned} 48b &= 3\left(8(c_2 - 2) + \frac{11}{8}\left(c_2^2 - 2c_2\right)\right) \\ &= 11c_1^2 + 2c_2 - 48. \end{aligned} \tag{16}$$

When $\sigma > 0$,

$$\begin{aligned} 48b &= 3\left(8(c_2 - 2) - 11\left(c_1^2 - 2c_2\right)\right) \\ &= -11c_1^2 + 43c_2 - 48. \end{aligned} \tag{17}$$

Adding the two inequalities gives $c_2 \geq \frac{32}{15}$. This inequality is satisfied for complex surfaces, since $c_1^2 \geq 0$ and $c_2 \geq 3$. However, for negative signature,

$$11c_1^2 + 2c_2 - 48 \geq -42. \tag{18}$$

For positive signature, $c_1^2 \leq 3c_2$ and

$$-11c_1^2 + 43c_2 - 48 \geq 10c_2 - 48 \geq -18. \tag{19}$$

It is clear that $48b = 3(8b_2 - 11|\sigma|)$ is integer. By Rohlin's theorem, the signature will be divisible by 16 and b also would be integer. By Eqs. (18) and (19), $b \geq -\frac{7}{8}$ and $b \geq -\frac{3}{8}$ respectively. Then $b \geq 0$ and the $\frac{11}{8}$ conjecture is valid for simply connected complex surfaces with an even intersection form.

3. Summary of the K-theoretic proof of the lesser lower bound for the second Betti number

The Dirac operator D is a map from sections of spinor bundles E_o to E^1, D: $\Gamma(E^0) \rightarrow (E^1)$, and *ind D = dim Ker D – dim Coker D*. Now consider a Whitney sum with a finite-dimensional vector bundle, such that $D = L + L' : V^0 \oplus \Gamma(E^0) \rightarrow V^1 \oplus \Gamma(E^1)$, where L is a finite-dimensional mapping and L' is an isomorphism between infinite-dimensional spaces. Then *ind D = dim* V_0 *– dim* V_1. Therefore, topological information about a manifold on which the Dirac operator, arising from equations with a linearization of $N = 2$ supersymmetry on the space, deduced from the index may be evaluated through a finite-dimensional construction. When M is a closed spin 4-manifold, the Seiberg-Witten map is a Pin_2-equivariant map given by $\mathbb{H}^\infty \oplus \tilde{R}^\infty \rightarrow \mathbb{H}^\infty \oplus \tilde{R}^\infty$, where $\tilde{R}$ is the nontrivial one-dimensional real representation space of Pin_2. A finite-dimensional approximation is a Pin_2-equivariant map $\mathbb{H}^{c_0} \oplus \tilde{R}^{d_0} \rightarrow \mathbb{H}^{c_1} \oplus \tilde{R}^{d_1}$, where $c_0 - c_1 = -\frac{\sigma(M)}{16}$ and $d_0 - d_1 = b^+(M)$, in a generalized Kuranshi construction [9].

The four-dimensional spin manifold will admit a $Spin_4$ bundle and vector bundles T, S^+, S^- and Λ constructed from the $Spin_4 \times Pin_2$ modules $-\mathbb{H}_+, {}_+\mathbb{H}, {}_-\mathbb{H}$ and ${}_+\mathbb{H}_+$ defined by the actions $q_-aq_+^{-1}, q_+\phi q_0^{-1}, q_+\omega q_0^{-1}$ and $q_+\omega q_+^{-1}$ for $(q_-, q_+, q_0) \in Spin_4 \times Pin_2$ and $a \in {}_-\mathbb{H}_+, \phi \in {}_+\mathbb{H}, \psi \in -\mathbb{H}$ and $\omega \in {}_+\mathbb{H}_+$. If $\tilde{R}$ is the real one-dimensional Pin_2 module defined by multiplication by Pin_2/S^1, $\tilde{T} = T \otimes \tilde{R}, C : T \otimes S^+ \rightarrow S^-$ with $(a, \phi) \rightarrow a\phi, \overline{C} : T \otimes \tilde{T} \rightarrow \tilde{\Lambda}$ with $(a, b) \rightarrow \overline{a}b, D_1 = C\nabla_1 : \Gamma(S^+) \rightarrow \Gamma(S^-), D_2 = \overline{C}\nabla_2 : \Gamma(\tilde{T}) \rightarrow \Gamma(\tilde{\Lambda}), D = D_1 \oplus D_2 : \Gamma(S^+ \oplus \tilde{T}) \rightarrow \Gamma(S^- \oplus \tilde{\Lambda}), Q : S^+ \oplus \tilde{T} \rightarrow S^- \oplus S^- \oplus \tilde{\Lambda}$ with $(\phi, a) \rightarrow (a\phi i, \phi i\overline{\phi})$, then $D + Q$: $V \rightarrow W$ where V is the L_4^2 completion of $\Gamma(S^+ \oplus \tilde{T})$ and W is the L_3^2 completion of $\Gamma(S^- \oplus \tilde{\Lambda})$ [4].

Let M be a compact G-space, E and F be G-equivariant complex vector bundles over M, BE and BF be disk bundles corresponding to E and F, SE and SF be boundary sphere bundles, $\tilde{f} : BE \rightarrow BF$ be a G-equivariant bundle map preserving boundaries. By the Thom isomorphism theorem, $K_G(BE, SE)$ and $K_G(BF, SF)$ are generated by Thom classes τ_E and τ_F. With $\tilde{f}^*$ being the pullback map $K_G(BF, SF) \rightarrow K_G(BE, SE)$, $\tilde{f}^* \tau_F = \alpha_0 \tau_E$, where $\alpha_0 \in K_G(M)$ is the degree of $\tilde{f}^*$. Since the restriction of the Thom classes to the zero sections are the Euler classes of E and F, $\sum_d (-1)^d \left[\Lambda^d F\right] = \alpha_0 \sum_d (-1)^d \left[\Lambda^d E\right]$.

Contracting $M = pt.$, $G = Pin_2$,

$$
\begin{aligned}
E_{pt.} &= V_{\lambda,\mathbb{C}} = \left(\mathbb{H}^{k+m} + \tilde{R}^n\right) \otimes \mathbb{C} \\
F_{pt.} &= \overline{W}_{\lambda,\mathbb{C}} = \left(\mathbb{H}^m + \tilde{R}^{b_+ + n}\right) \otimes \mathbb{C},
\end{aligned}
\tag{20}
$$

where V_λ is the subspace of V spanned by the eigenspace of $D^* D$ with eigenvalues less than or equal to λ, and W_λ is the subspace of w spanned by eigenspaces of DD^* with eigenvalues less than or equal to λ,$f : V_{\lambda,\mathbb{C}} \rightarrow \overline{W}_{\lambda,\mathbb{C}}$ is the complexification of $D_\lambda + Q_\lambda$,where $D_\lambda + Q_\lambda = (D + p_\lambda Q)\big|_{V_\lambda}$,$f(u \otimes 1 + v \otimes i) = (D_\lambda + Q_\lambda)u \otimes 1 + (D_\lambda + Q_\lambda)v \otimes i$.

Suppose that $\varphi : V = ker(d^*) \oplus \Gamma(V^+) \rightarrow \Omega_+^2 \oplus \Gamma(V^-) = W$, $\varphi(v) = L(v) + \theta(v)$, $L = \begin{pmatrix} d^+ & 0 \\ 0 & \partial \end{pmatrix}$ is linear, $\theta(a, \psi) = (\sigma(\psi), a\psi)$ is quadratic, where a is the gauge potential in the covariant derivative and σ' is an automorphism of the space $\Gamma(W^+)$ and $\sigma'\left(\psi_j\right) = \sigma'(z + jw)j = \sigma'(-\bar{v} + j\bar{m}z) = -\sigma'(\psi)$. Let $f_\lambda : V \rightarrow W$ be defined by $u \equiv f_\lambda(v) = v + L^{-1}(1 - p_\lambda)\theta(v)$, $L(u) = L(v) + (1 - p_\lambda)\theta(v)$, with p_λ being the projection of V and W onto V_λ and W_λ. Defining $\varphi_\Lambda : \oplus_{\lambda \leq \Lambda} V_\lambda \rightarrow \oplus_{\lambda \leq \Lambda} W_\lambda$, $\varphi_\Lambda(u) = p_\lambda \varphi f_\Lambda^{-1}(u)$ [10].

Let $T_u(v) = u - L^{-1}(1 - p_\Lambda)\theta(v)$. Then

$$
\begin{aligned}
\|T_u(v_1) - T_u(v_2)\| &= \|(T_u(v_1) - u) - (T_u(v_2) - u)\| \\
&= \left\|-L^{-1}(1 - p_\Lambda)\theta(v_1) + L^{-1}(1 - p_\Lambda)\theta(v_2)\right\|.
\end{aligned}
\tag{21}
$$

The eigenvalue of $L^{-1}(1 - p_\Lambda)$ is $\frac{1}{\lambda}$ on each W_λ, which has the maximum value $\frac{1}{\Lambda}$ for $\lambda > \Lambda$. The automorphism $\sigma'(\psi)$ is given by $\sigma'(z, w) = i\left(\frac{|z|^2 - |w|^2}{2}\right) - k\, Re\,(z\overline{w}) + jIm(z\overline{w})$, and, if $\hat{\sigma}(z, w) = \hat{\rho} \circ \sigma'(z, w)$, where $\hat{\rho} : T^*(X) \rightarrow Hom\left(W^\pm, W^\mp\right)$, $\hat{\rho}(v_1 \wedge v_2) = \frac{1}{2}[\rho(v_1), \rho(v_2)]$,$f_1 = \frac{1}{2}(e^1 \wedge e^2 \pm e^3 \wedge e^4)$,$f_2 = \frac{1}{2}(e^1 \wedge e^3 \pm e^4 \wedge e^2)$,$f_3 = \frac{1}{2}(e^1 \wedge e^4 \pm e^2 \wedge e^3)$, where i, j, k correspond to $f_1, f_2, f_3 \in \Lambda^+(M)$, with $\Lambda^2(M) = \Lambda^+(M) \otimes \Lambda^{-1}(M)$, $|\hat{\sigma}(\psi)|^2 = \frac{1}{2}|\psi|^2$ [3]. It follows that

$$
\left\|(\hat{\sigma}\psi_1,\ a_1\psi)_1) - (\hat{\sigma}\psi_2, a_2\psi_2)\right\| \leq \|(a_1,\ \psi_1) - (a_2, \psi_2)\|
\tag{22}
$$

if $|\psi_1| < a_1$, $|\psi_2| < a_2$, and $a_1, a_2 < 1$. Under these conditions, by the Banach contraction principle, $\varphi_\Lambda^{-1}(0)$ is a compact set.

With $BV_{\lambda,\mathbb{C}} = \{u \otimes 1 + v \otimes i \in V_{\lambda,\mathbb{C}} \| \|u\|, \|v\| \leq R\}$, $SV_{\lambda,\mathbb{C}} = \partial BV_{\lambda,\mathbb{C}}$, let $\tilde{f} = f \circ p$, where $p : \overline{W}_{\lambda,\mathbb{C}} \backslash \{0\} \to S\overline{W}_{\lambda,\mathbb{C}}$. Then the mapping $\tilde{f} : BV_{\lambda,\mathbb{C}} \to B\overline{W}_{\lambda,\mathbb{C}}$ is defined to be the cone of $\tilde{f}$. If $k > 0, \{\alpha \in R(Pin_2) | \rho^\ell(F)\alpha = (\psi^\ell \alpha)\rho^\ell(E)\} \subset Ker(R(Pin_2) \to R(S^{-1}))$. Consider an element α of $Ker(R(Pin_2) \to R(S^1))$ satisfying $\sum_d (-1)^d [\Lambda^d F] = \alpha \sum_d (-1)^d [\Lambda^d E]$. Regarding E and F as S^1 modules, let $E' = 2(k+m)(\mathbb{C} \oplus \mathbb{C}^*) \oplus n$ and $F' = 2m(\mathbb{C} \oplus \mathbb{C}^*) \oplus (b_+ + n)$ be representation spaces.

Let ψ^ℓ be the Adams operation and $\rho^\ell(E)$ be the characteristic class satisfying $\psi^\ell \tau_E = \rho^\ell(E)\tau_E$. Then

$$\begin{aligned}
\psi^\ell\left(\tilde{f}^* \tau_F\right) &= \tilde{f}^* (\psi^\ell \tau_F) = \tilde{f}^* (\rho^\ell(F)\tau_F) \\
\psi^\ell(\alpha_0 \tau_E) &= (\psi^\ell \alpha_0)\psi^\ell \tau_E = \psi^\ell \alpha_0 \rho^\ell(E)\tau_E \\
\tilde{f}^* (\rho^\ell(F)\tau_F) &= \rho^\ell(F)\tilde{f}^* \tau_F = \rho^\ell(F)\alpha_0 \tau_E \\
\rho^\ell(F)\alpha_0 \tau_E &= (\psi^\ell \alpha_0)\rho^\ell(E)\tau_E \\
\rho^\ell(F)\alpha_0 &= (\psi^\ell \alpha_0)\rho^\ell(E).
\end{aligned} \tag{23}$$

Given that $\rho^\ell(L) = 1 + [L] + [L^2] + \cdots + [L^{\ell-1}]$ for a line bundle L, $\rho^\ell(E') = \{\rho^\ell(\mathbb{C})\rho^\ell(\mathbb{C}^*)\}^{2(k+m)}\rho^\ell(1)^n = \{(1 + t + \ldots + t^{\ell-1})(1 + t^{-1} + \ldots + t^{-(\ell-1)})\}^{2m} \ell^n$ and $\rho^\ell(F') = \{(1 + t + \ldots + t^{\ell-1})(1 + t^{-1} + \ldots + t^{-(\ell-1)})\}^{2m} \ell^{b_+ + n}$. Since $Ker(R(Pin_2) \to R(S^{-1})) = \{c(1 - \tilde{1}) | c \in \mathbb{Z}\}$, the trace of the degree relation gives $2^{2m + b_+ + n} = 2c\, 2^{2k + 2m + n}$ for $\alpha = c(1 - \tilde{1})$, which is consistent with the inequality $b_+ \geq 2k + 1$.

Methods have been developed for increasing the bound for $\frac{b_2}{|\sigma|}$ through the inequality between $b_2(M)$ and the level number of $\mathbb{CP}^{2k-1}$, defined to be least n such that $[\mathbb{CP}^{2k-1}, S^{n-1}]^{\mathbb{Z}_2} \neq \emptyset$, where $k = -\frac{\sigma(M)}{16}$ [11, 12]. The computations of level($\mathbb{CP}^{2k-1}$) yield the inequalities $level((\mathbb{CP}^{2k-1}) \geq 2k + t$ if $k \equiv t (mod\, 4), t = 1, 2, 3$ and the equality $level(\mathbb{CP}^{2k-1}) = 2k + 3$ if $k \equiv 0\ (mod\, 4), k > 0$ [13]. The equivalent inequalities for the second Betti number and signature would be $b_2 \geq \frac{5}{4}|\sigma| + 2t$ for $\frac{|\sigma|}{16} \equiv t\,(mod\, 4)$ when $t = 1, 2, 3$ and $b_2 \geq \frac{5}{4}|\sigma| + 6$ for $\frac{|\sigma|}{16} \equiv 0\,(mod\, 4), |\sigma| > 0$. The Bauer-Furuta stable Seiberg-Witten invariants also yield a condition for the existence of smooth structures [14].

4. Proof of the exact bound for the second Betti number

It will be demonstrated that the previous lower bound for b_+ can be increased to a maximal value.

Theorem 4.1. The second Betti number and signature of a smooth, oriented, simply connected, compact, spin four-manifold with an indefinite intersection form satisfies the inequality $b_2 \geq \frac{11}{8} \geq |\sigma|$.

Proof. Since c is a non-zero integer, the trace condition also requires a stricter bound for b_+. The Euler class of E will be given by that of $\overline{\mathbb{H}}^{k+m}$, while the Euler class of

F would be that of $\overline{\mathbb{H}}^m$, since the $\tilde{R}$ components do not contribute. The $Spin_4 \times Pin_2$ actions do not alter the norm of the points in the quaternionic vector spaces. The fixed point at the origin, however, would be the source of a flow generated by a dilation, representing an invariance of the spinor equation, that must have an endpoint at ∞. Adding the point $\{\infty\}$ to the quaternionic plane produces a manifold that is diffeomorphic to S^4 and $_{\chi}(S^4) = 2$. Therefore, by the Thom class relation,

$$2^m = tr(\alpha_{pt.})2^{k+m} \tag{24}$$

and $tr(\alpha_{pt.}) = 2^{-|k|}$, except that k must be chosen to be nonpositive through $k = -|k|$, then $tr(\alpha_{pt.}) = 2^{|k|}$. A similar result follows from the evaluation of the K-theory characteristic classes $\rho^{\ell}(E)$. Reintroducing the $\tilde{R}$ components in the vector bundles allows the inclusion of the factor $c_0 \geq 1$ occurring in $Ker(R(Pin_2) \to R(Pin_1))$ in $\alpha = 2^{|k|-1}c_0(1-\tilde{1})$.It follows that

$$2^{2m+b_+ +n} = 2c_0\, 2^{|k|-1} 2^{2|k|+2m+n} = 2^{|k|}2^{2|k|+2m+n} \tag{25}$$

or

$$b_+ \geq 3|k| \tag{26}$$

and

$$b_2 \geq \frac{11}{8}|\sigma|. \tag{27}$$

Furthermore, this inequality is equivalent to $n \geq \frac{3}{2}|m|$ since the second Betti number and the absolute value of the signature of a manifold with the intersection form $mE_8 \oplus n\begin{pmatrix} 0 & 1 \\ 1 & 0 \end{pmatrix}$ are $b_2 = 8|m| + 2n$ and $|\sigma| = 8|m|$, and

$$\begin{aligned} \frac{b_2}{|\sigma|} = \frac{8|m| + 2n}{8|m|} &\geq \frac{11}{8} \\ 8|m| + 2n &\geq 11|m| \\ n &\geq \frac{3}{2}|m|. \end{aligned} \tag{28}$$

It has been proven for cobordisms between homology three-spheres Y_0 and Y_1 with the intersection form $m(-E_8) + n\begin{pmatrix} 0 & 1 \\ 1 & 0 \end{pmatrix}$ that $\kappa(Y_1) + n \geq \kappa(Y_0) + m + 1$, where κ is an invariant that reduces modulo 2 to the Rohlin invariant $\mu(Y)$ [2]. When Y_0 and Y_1 are S^3, this inequality is $n \geq |m| + 1$, which is consistent with the previously derived inequality [4] for the coefficients since it would follow that

$$\begin{aligned} &\frac{b_2}{|\sigma|} = \frac{8|m| + 2n}{8|m|} \geq \frac{5}{4} + \frac{2}{|\sigma|} \\ &8|m| + 2n \geq 10|m| + 2 \\ &n \geq |m| + 1. \end{aligned} \tag{29}$$

Nevertheless, the nonexistence of smooth spin manifolds transcending the stricter inequality, such that $n = \frac{3}{2}|m| - 1$, with a decomposition $M = X_1 \cup_{Y_1} X_2 \cup_{Y_2} \dots \cup_{Y_{r-1}} X_r$, where the intersection forms of X_i, $1 \leq i \leq r-1$ are $2(-E_8) + 3\begin{pmatrix} 0 & 1 \\ 1 & 0 \end{pmatrix}$, Y_i is a homology three-sphere and X_r has the intersection form $2(-E_8) + 2\begin{pmatrix} 0 & 1 \\ 1 & 0 \end{pmatrix}$ [15], indicates there are characteristics that cannot be preserved when under topological sums yielding $\frac{5}{4} < \frac{b_2}{|\sigma|} < \frac{11}{8}$.

Definition 4.2. A space is Floer G-split if the S^1 action on the the K-theory group $\tilde{K}_G(M)$ has the form $z \to z^r$ for some positive integer r.

Theorem 4.3. The compact manifold $M = X_1 \cup_{Y_1} X_2 \cup_{Y_2} \dots \cup_{Y_{r-1}} X_r$ with spaces X_i, = 1, … , $r-1$, and X_r having the intersection forms $Q(X_i) = 2(-E_8) \cup 3H$ and $Q(X_r) = 2(-E_8) \cup q_r H$, and Y_i being homology three-spheres, is smooth if and only if $q_r \geq 3$.

Proof. When the S^1 action does not have this form, which requires $z \to z^r$ with $r \notin \mathbb{Q}$, the resulting quotient produces a non-Hausdorff structure because arbitrarily near points are identified. If $r = \frac{s}{t} \in \mathbb{Q} \backslash \mathbb{Z}$, then the points $e^{2\pi i\theta}$ and $e^{\frac{2\pi}{t} i\theta}$ are identified, which yields an orbifold rather than a smooth manifold. A theorem proven by Manolescu states that no closed spin four-manifold has a decomposition of this type such that all homology spheres in the set $\{Y_i\}$ are Floer G-split [2]. It follows that a quotient of one of the homology spheres by an S^1 action with $z \to z^r, r \notin Z$, is not smooth. The S^1 action on the homology sphere, which is not Floer G-split, may be transferred to the manifold X_r, as a result of the parallelizability necessary for the existence of the spin structure, thereby proving that a spin manifold with an intersection form $2(-E_8) \oplus 2H$ does not admit a smooth structure.

Now suppose that the spin manifold X has the intersection form $m(-E_8) \oplus nH$, where $n = \frac{3}{2}|m|$. The analogous result to that given above for $n = \frac{3}{2}|m| - 1$ would be the decomposition

$$\begin{aligned} X &= X_1 \cup_{Y_1} X_2 \cup_{Y_2} \dots \cup_{Y_{r-1}} X_r \\ Q(X_i) &= 2(-E_8) + 3H, i = 1, \dots, r \end{aligned} \tag{30}$$

with $Y_1, \dots, Y_{r-1}$ being homology spheres, then the inequality $\kappa(Y_{i+1}) + 3 \geq \kappa(Y_i) + 2 + 1$, or $\kappa(Y_{i+1}) \geq \kappa(Y_i)$, is valid for all $j = 1, \dots, r-2$, and each Y_i is Floer G-split, which requires the existence of a smooth structure on each X_i, $i = 1, \dots, r$. Therefore, closed spin four-manifolds with the intersection form $mE_8 \oplus nH$ and $n = \frac{3}{2}|m|$ admit smooth structures. For $n \geq \frac{3}{2}|m|$, the inequalities for $\kappa(Y_i)$, $i = 1, \dots, r-1$ continue to be valid, each of the homology spheres will be Floer K_G-split, and there will be a smooth structure on the spin four-manifold.

Several results may be proven given the validity of the $\frac{11}{8}$ conjecture, including the theorem on $\xi \cdot \xi$ for a characteristic second homology class ξ representable by S^2 for a range of values of b_+ and b_- [7]. The following lemma is required:

Let M be a closed connected oriented four-manifold with $\xi \in H_2(M; \mathbb{Z})$ be a characteristic homology class representable by S^2. Then $\xi \cdot \xi = \sigma(M) + 16\,m$ with $m \leq \max\left\{ \lfloor \frac{b_1 - 1}{3} \rfloor, \lfloor \frac{b_- - b_+}{16} \rfloor \right\}$ would be consistent with the $\frac{11}{8}$ conjecture.

The demonstration of this result is suggestive of an equivalence of the conditions with the limits of m being derived from geometric properties of the spin manifold, and the $\frac{11}{8}$ conjecture following from the ranges for m.

By a theorem of Kervaire and Milnor, it is known that, in a four-manifold which allows the embedding of two-spheres representing the homology class ξ, $\xi \cdot \xi = \sigma(M) + 16\,m$ for some m [16]. Since $\sigma(M) = b_+ - b_-$, $\xi \cdot \xi = b_+ - b_- + 16\,m$. The range given in the above theorem with $\xi \cdot \xi = 0$ or $b_- = b_+ + 16\,m$ yields the following results.

$$m \le max\left\{\lfloor\frac{b_+ - 1}{3}\rfloor, \lfloor\frac{b_+ + 16m - r}{16}\rfloor\right\} = max\left\{\lfloor\frac{b_+ - 1}{3}\rfloor, m + \lfloor\frac{b_+ - r}{16}\rfloor\right\}$$
$$= max\left\{\lfloor\frac{b_+ - 1}{3}\rfloor, m\right\} \tag{31}$$

and

$$m \le \lfloor\frac{b_+ - 1}{3}\rfloor. \tag{32}$$

Since spin manifolds have an even intersection form, b_+ is even. Then if $b_+ = 0, 2$ *or* 4 (*mod* 6) and $\lfloor\frac{b_+-1}{3}\rfloor = \lfloor\frac{b_+}{3}\rfloor - 1, \lfloor\frac{b_+}{3}\rfloor\ or \lfloor\frac{b_+}{3}\rfloor$ respectively. If $m \le \frac{b_+}{3}$,

$$b_- = b_+ + 16m \le \frac{19}{3}b_+. \tag{33}$$

Then

$$\begin{aligned} b_2 - \sigma &\le \frac{19}{3}(b_2 + \sigma) \\ \frac{16}{3}b_2 &\ge -\frac{22}{3}\sigma \\ b_2 &\ge \frac{11}{8}|\sigma| \sigma < 0. \end{aligned} \tag{34}$$

For $\sigma > 0$, the roles of b_+ and b_+ are interchanged, and

$$\begin{aligned} b_2 + \sigma &\le \frac{19}{3}(b_2 - \sigma) \\ \frac{16}{3}b_2 &\ge \frac{22}{3}\sigma \\ b_2 &\ge \frac{11}{8}|\sigma| \sigma > 0. \end{aligned} \tag{35}$$

Therefore, the condition $\xi \cdot \xi = 0$ together with the range of m yielding the upper limit $\frac{b_+}{3}$, is sufficient to prove the $\frac{11}{8}$ conjecture. Given that $\xi \cdot \eta$ is the intersection number of ξ and η are representable by S^2, $\xi \cdot \xi$ would equal the sum of the eigenvalues of the intersection form of $S^2 \times S^2$, which equals zero.

The connected sum $\mathbb{CP}^2\#9\overline{\mathbb{CP}^2}$ does not satisfy the inequalities in Eq. (14) for the coefficients k and ℓ. Nevertheless, it has exotic smooth structures. The nonexistence of spin structures on $\mathbb{CP}^2\#9\overline{\mathbb{CP}^2}$ may be demonstrated [17, 18]. The spaces $\mathbb{CP}^2\#\ell\overline{\mathbb{CP}^2}$ have $\frac{b_2}{|\sigma|} < \frac{11}{8}$ for $\ell \ge 7$. Furthermore, these four-manifolds have both standard and exotic smooth structures for $k = 7$, 8 and 9 [19–21].

Proposition 4.4. The topological sums $\mathbb{CP}^2\#\ell\overline{\mathbb{CP}^2}, \ell \geq 7$, do not represent counterexamples to the inequality $\frac{b_2}{|\sigma|} \geq \frac{11}{8}$ required for smooth structures on spin manifolds given the validity of the $\frac{10}{8}$ theorem.

Proof. The existence of smoooth structures on these spaces is established. From §2, the second Betti number and signature of $\mathbb{CP}^2\#\ell\overline{\mathbb{CP}^2}$ equal

$$\begin{aligned} b_2 &= 1 + \ell \\ \sigma &= 1 - \ell. \end{aligned} \tag{36}$$

Then

$$\frac{b_2}{|\sigma|} = \frac{\ell + 1}{\ell - 1} \quad \ell \geq 2. \tag{37}$$

For $\ell = 7$, 8 and 9, $\frac{b_2}{|\sigma|}$ is $\frac{4}{3}, \frac{9}{7}$ and $\frac{5}{4}$ respectively. Then

$$\begin{aligned} &\frac{10}{8} < \frac{b_2}{|\sigma|} < \frac{11}{8} \quad for\, \ell = 7{,}8 \\ &\frac{b_2}{|\sigma|} = \frac{10}{8} \quad for\, \ell = 9 \\ &\frac{b_2}{|\sigma|} < \frac{10}{8} \quad for\, \ell > 9. \end{aligned} \tag{38}$$

Consequently, the values $\ell \geq 9$ must be covered by the $\frac{10}{8}$ theorem, which will require the absence of spin structures on these manifolds.

It follows from Rohlin's theorem that the signature of a smooth, spin compact four-manifold is divisible by 16. For $\mathbb{CP}^2\#\ell\overline{\mathbb{CP}^2}$, this condition is

$$\ell \equiv 1 \ (mod\ 16). \tag{39}$$

This congruenece condition is not satisfied by $\ell = 7$ or $\ell = 8$. Therefore, there will be no spin structure for these values. It follows that the connected sums for $\ell \geq 7$ will not represent a counterexample to the $\frac{11}{8}$ conjecture when the lower bound of $\frac{10}{8}$ suffices generally for smooth spin manifolds.

There would be a spin structure on $\mathbb{CP}^2\#\ell\overline{\mathbb{CP}^2}$. Both the above proposition and the consistency of the $\frac{10}{8}$ theorem require the nonexistence of spin structures on $\mathbb{CP}^2\#(16r + 1)\overline{\mathbb{CP}^2}$ for $r \geq 1$.

Proposition 4.5. The topological sum $\mathbb{CP}^2\#(16r + 1)\overline{\mathbb{CP}^2}$ is a spin manifold only if $r = 0$.

Proof. There exists a spin structure on a space M the second Stiefel-Whitney class $w_2(M) \in H_2(M; \mathbb{Z}_2)$ is nonvanishing. The second homology group of a connected sum $M_1\#M_2$, *where*dim M_1 = *dim* M_2 = 4, is

$$H_2(M_1\#M_2) = H_2(M_1) \oplus H_2(M_2) \tag{40}$$

and. Specializing to the group $\mathbb{Z}_2$,

$$H_2(M_1\#M_2, \mathbb{Z}_2) = H_2(M_1, \mathbb{Z}_2) \oplus H_2(M_2, \mathbb{Z}_2). \tag{41}$$

Then

$$H_2\left(\mathbb{CP}^2\#\overline{\mathbb{CP}^2}, \mathbb{Z}_2\right) = H_2(\mathbb{CP}^2, \mathbb{Z}_2) \oplus H_2\left(\overline{\mathbb{CP}^2}, \mathbb{Z}_2\right) \tag{42}$$

Since $\mathbb{CP}^2$ does not have a spin structure, there will be a nonvanishing generator of the second homology group $[u] \neq [0]$ and

$$H_2\left(\mathbb{CP}^2\#\overline{\mathbb{CP}}^2, \mathbb{Z}_2\right) = [u] \oplus - [u] \tag{43}$$

The element of the homology group must be an element of $\mathbb{Z}_2$. Therefore, it would be the image of the $[u] \oplus - [u]$ under the mapping

$$\varphi : H_2(\mathbb{CP}^2, \mathbb{Z}_2) \oplus H_2\left(\overline{\mathbb{CP}}^2, \mathbb{Z}_2\right) \to \mathbb{Z}_2 \tag{44}$$

This homomorphism will be defined by

$$\varphi([u_1] \oplus [u_2]) = [u_1] + [u_2] \in \mathbb{Z}_2 \tag{45}$$

Since

$$\varphi([u] \oplus - [u]) = [u] - [u] = [0], \tag{46}$$

$H_2\left(\mathbb{CP}^2\#\overline{\mathbb{CP}^2}, \mathbb{Z}_2\right) = \emptyset$, the second Stiefel-Whitney class vanishes, and there is a spin structure for $r = 0$.

The homology group for higher values of r equals

$$H_2\left(\mathbb{CP}^2\#(16r+1)\overline{\mathbb{CP}^2}\right) = H_2(\mathbb{CP}^2)\#(16r+1)H_2\left(\overline{\mathbb{CP}^2}, \mathbb{Z}_2\right). \tag{47}$$

Given that $[u] \in H_2(\mathbb{CP}^2, \mathbb{Z}_2)$ and $-[u] \in H_2\left(\overline{\mathbb{CP}^2}, \mathbb{Z}_2\right)$, the element of $\mathbb{Z}_2$ for the topological sum is

$$\varphi([u] \oplus (16r+1)(-u)) = -16r[u] \tag{48}$$

Multiplication by a non-zero scalar does not affect the generator of the nontrivial second homology class, which does not vanish. Then, $\mathbb{CP}^2\#(16r+1)\overline{\mathbb{CP}^2}$ is not a spin manifold for $r > 1$.

There are no counterexamples given by $\mathbb{CP}^2\#\overline{\mathbb{CP}^2}$ to the lower bound of $\frac{11}{8}$ for $\frac{b_2}{|\sigma|}$. Topological sums with S^4, $S^2 \times S^2$ and $K3$ will not affect this inequality for the ratio of the second Betti number to the magnitude of the signature. No other potential counterexamples can exist for smooth, simply connected, compact spin four-manifolds.

5. The local coefficients for manifolds with a spin covering

The proof in §4 is restricted to smooth, oriented, simply connected, compact manifolds which admit spinor structures. It remains to be established if the conclusions continue to be valid for smooth non-spin four-manifolds that have a spin covering. Since the lower bound for $\frac{b_2}{|\sigma|}$ has been increased to $\frac{11}{8}$.

Theorem 5.1. The coefficients of the intersection form $mE_8 + nH$, where $H = \begin{pmatrix} 0 & 1 \\ 1 & 0 \end{pmatrix}$ satisfy the inequality $n \geq \frac{3}{2}m - \rho(M) + \frac{\rho(M)}{2^i}$, with i being the exponent in the order of a spin covering of the smooth, oriented, simply connected, compact manifold M and $\rho(M)$ is the rank of $H_1(M; \mathbb{Z}_2)$.

Proof. Let M be a smooth, oriented, simply connected, compact four-manifold with the even intersection form $I = mE_8 + nH$, signature $\sigma(M) = 8\,m$ and Euler number $e(M) = 2 + 8\,m + 2n$, since the first Betti number can be set equal to zero for a given intersection form. There exists a 2^i cyclic covering $\pi: N \to M$, where N is a smooth, oriented spin manifold [22, 23]. The signature and the Euler number of the covering space are

$$\begin{aligned} \sigma(N) &= 2^i\sigma(M) = 8\left(2^i m\right) \equiv 8r \\ e(N) &= 2^i e(M) = 2^{i+1} + 8\left(2^i m\right) + 2\left(2^i n\right) = 2 + 8r + 2s \end{aligned} \tag{49}$$

then $N = rE8 + sH$, where

$$r = 2^i ms = 2^i n + 2^i - 1. \tag{50}$$

Since it has been proven that $s \geq |r| + 1$ for spin manifolds [4],

$$2^i n + 2^i - 1 = 2^i|m| + 1. \tag{51}$$

and

$$n \geq |m| - \left(1 - \frac{1}{2^{i-1}}\right). \tag{52}$$

A term $b_1(N)$ can be added to s to give $s + b_1(N) \geq r + 1$. Since $b_1(N) \leq (2^i - 1)(\rho(M) - 1)$ when $b_1(M) = 0$, where $\rho(M)$ is the rank of $H_1(M; \mathbb{Z}_2)$ [6]. Then

$$\begin{aligned} 2^i n + 2^i - 1 + \left(2^i - 1\right)(\rho(M) - 1) &\geq 2^i|m| + 1. \\ n \geq |m| - \rho(M) + \frac{1 + \rho(M)}{2^i}&. \end{aligned} \tag{53}$$

If $\rho(M) = 1$ and $N \to M$ is a double covering with $H_1(M; \mathbb{Z}) = \mathbb{Z}_2$ [6], the inequality $n \geq |m|$ is valid.

With a tighter bound $n \geq \frac{3}{2}|m|$ for spin manifolds, a similar inequality will be found for non-spin manifolds. By the proof in §4, $s \geq \frac{3}{2}|r|$ for the spin covering N, or equivalently,

$$\begin{aligned} 2^i n + 2^i - 1 &\geq \frac{3}{2}2^i|m| \\ n &\geq \frac{3}{2}|m| - \left(1 - \frac{1}{2^i}\right). \end{aligned} \tag{54}$$

Since $(2^i-1)(\rho(M)-1)$ is an upper bound for $b_1(N)-b_1(M)$, after $b_1(M)$ is set equal to zero,

$$2^i n + \left(2^i - 1\right)\rho(M) = \frac{3}{2}2^i|m| \tag{55}$$

or

$$n \geq \frac{3}{2}|m| - \rho(M) + \frac{\rho(M)}{2^i}. \tag{56}$$

It has been proven that there exist nonsmoothable spin manifolds with $b_2 \geq \frac{5}{4}|\sigma| + 2$ [24]. The strict inequality yields a contradiction with the demarcation between smooth and non-smooth structures on a spin four-manifold, which conjectured for coefficients of the intersection form generally.

Similarly, it is claimed that there are nonsmoothable non-spin manifolds with $b_2 \geq \frac{5}{4}|\sigma|$. The inequality derived for non-spin manifolds $n \geq |m| - \left(1 - \frac{1}{2^{i-1}}\right)$ may be translated to a bound for the second Betti number.

$$\begin{aligned} \frac{b_2}{|\sigma|} &= \frac{8|m| + 2n}{8|m|} \\ &\geq 1 - \frac{1}{4}\left(1 + \frac{1}{|m|}\left(1 - \frac{1}{2^{i-1}}\right)\right) \\ &= \frac{5}{4} - \frac{2}{|\sigma|}\left(1 - \frac{1}{2^{i-1}}\right). \end{aligned} \tag{57}$$

This lower bound for b_2 is less than or equal to $\frac{5}{4}|\sigma|$ for $i \geq 1$.

The tighter inequality for the coefficients in the intersection form is equivalent to

$$\begin{aligned} \frac{b_2}{|\sigma|} &\geq 1 + \frac{1}{4}\left(\frac{3}{2} - \frac{1}{|m|}\left(1 - \frac{1}{2^i}\right)\right) \\ &= \frac{11}{8} - \frac{2}{|\sigma|}\left(1 - \frac{1}{2^i}\right). \end{aligned} \tag{58}$$

The contradiction is resolved in the inequality if the lower bound for $\frac{b_2}{|\sigma|}$ can be increased. Then, the nonsmooth manifolds can exist in the region $\frac{5}{4}|\sigma| + 2 \leq b_2 \leq \frac{11}{8}|\sigma|$ when a spin structure exists and $\frac{5}{4}|\sigma| - 2\left(1 - \frac{1}{2^{i-1}}\right) \leq b_2 \leq \frac{11}{8}|\sigma| - 2\left(1 - \frac{1}{2^i}\right)$ when there is no spin structure.

A lower bound for b_2 also can be derived for smooth, oriented non-spin manifolds by the following set of equations

$$\begin{aligned} dim_{\mathbb{Z}_2} H^2(M;\mathbb{Z}_2) &= b_2(M) + 2t \\ t &= dim_{\mathbb{Z}_2}\left(Tor_2 H_1(M;\mathbb{Z}_2 \times \mathbb{Z}_2)\right), \end{aligned} \tag{59}$$

$$b_1(N;\mathbb{Z}_2) \leq 2^i b_1(M;\mathbb{Z}_2) - 2^i + 1, \tag{60}$$

and

$$\begin{aligned} b_2(N;\mathbb{Z}_2) &= 2^i e(M) - 2 + 2b_1(N;\mathbb{Z}_2) \\ &\leq 2^i(2 - 2b_1(M;\mathbb{Z}_2) + b_2(M;\mathbb{Z}_2)) + 2\left(2^i b_1(M;\mathbb{Z}_2) - 2^i + 1\right) \\ &= 2^{i+1} - 2^{i+1}b_1(M;\mathbb{Z}_2) + 2^i b_2(M;\mathbb{Z}_2) + 2^{i+1}b_1(M;\mathbb{Z}_2) - 2^{i+1} + 2 \\ &= 2^i b_2(M;\mathbb{Z}_2) + 2, \end{aligned} \tag{61}$$

and, since the degree of the spin covering of M is even, t equals one [6]. It follows that

$$b_2(N) + 2 \leq 2^i(b_2(M) + 2) + 2 = 2^i b_2(M) + 2^{i+1} + 2 \tag{62}$$

or

$$b_2(N) \leq 2^i b_2(M) + 2^{i+1}. \tag{63}$$

By the strong $\frac{10}{8}$ inequality for spin manifolds,

$$b_2(N) \geq \frac{5}{4}|\sigma(N)| + 2 = \frac{5}{4}2^i|\sigma(M)| + 2. \tag{64}$$

and

$$\begin{aligned} b_2(M) &\geq \frac{1}{2^i}\left(\frac{5}{4}2^i|\sigma(M)| + 2 - 2^{i+1}\right) \\ &= \frac{5}{4}|\sigma(M)| - \left(2 - \frac{1}{2^{i-1}}\right). \end{aligned} \tag{65}$$

The tighter inequality derived in §4 for spin manifolds yields

$$b_2(N) \geq \frac{11}{8}|\sigma(N)| \geq \frac{11}{8}2^i|\sigma(M)| \tag{66}$$

and

$$\begin{aligned} b_2(M) &\geq \frac{1}{2^i}\left(\frac{11}{8}2^i|\sigma(M)| - 2^{i+1}\right) \\ &= \frac{11}{8}|\sigma(M)| - 2. \end{aligned} \tag{67}$$

The range for b_2 is narrower for $i \geq 1$ and there is a region below $\frac{11}{8}|\sigma|$ for which the existence of smooth structures remains to be established.

Theorem 5.2. An oriented, simply connected, compact four-manifold with an indefinite intersection form and a spin covering space has a smooth structure only if $\frac{b_2}{|\sigma|} \geq \frac{11}{8}$.

Proof. Consider an oriented, simply connected, compact, four-dimensional manifold M and the 2^i-fold spin covering $N \to M$. From the equation $\left(\tilde{f}^*\right)\tau_F = \alpha_0\tau_E$, where α_0 is the degree of the pull-back map $\tilde{f}^*$ from $K_G(BF, SF)$ to $K_G(BE, SE)$, the trace of the relation $\rho_\ell\left(\tilde{f}^*\right)\tau_F = (\psi^\ell\alpha_0)\rho^\ell(E)\tau_E$ when projected to an S^1 module in the subspace E' and F', introduces a factor of 2^i in the pull-back of the kernel of the map from Pin_2 to S^1, $\left\{\left(c\left(\mathbf{1} - \tilde{\mathbf{1}}\right)\middle| c \in \mathbb{Z}\right)\right\}^{2^i}$, and an overall factor of 2^{2^i}. Then,

$$\alpha = 2^{2^i} 2^{|k|} \tag{68}$$

and

$$b_+(N) \geq 2|k| + k + 2^i = 3|k| + 2^i. \tag{69}$$

The lower bound for the second Betti number of the spin covering N would be

$$b_2(N) = \frac{11}{8}|\sigma(N)| + 2^{i+1}. \tag{70}$$

By Eq. (63),

$$\begin{aligned} 2^i b_2(M) + 2^{i+1} &= \frac{11}{8}|\sigma(N)| + 2^{i+1} = \frac{11}{8} 2^i |\sigma(N)| + 2^{i+1} \\ b_2(M) &= \frac{11}{8}|\sigma(M)|. \end{aligned} \tag{71}$$

Therefore, the theoretically predicted inequality for the second Betti number of smooth, oriented four-manifolds has been derived.

6. Lower bound for the genus of a surface embedded in a four-manifold

The genus of an embedded surface Σ in a four-manifold M may be given the lower bound

$$g(\Sigma) > \left| \frac{2^r + 1}{2^{r+1}} \sum^2 - \frac{2^{r+1}}{2^r - 1}\sigma(M) \right| - \frac{2^{r-1}}{2^r - 1} b_2(M). \tag{72}$$

where Σ^2 is the intersection product of the second cohomology class Σ [19]. If the $\frac{11}{8}$ conjecture is true, the bound can be increased to

$$g(\Sigma) > \frac{11}{8}\left| \frac{2^r + 1}{2^{r+1}} \Sigma^2 - \frac{2^{r+1}}{2^r - 1}\sigma(M) \right| - \frac{2^{r-1}}{2^r - 1} b_2(M). \tag{73}$$

For an algebraic surface Σ_d of even degree d embedded in $\mathbb{CP}^2$, with $2\|d, g(\sum_d) \geq \frac{11}{32}d^2 - \frac{19}{9}$ [25].

The Thom conjecture for curves of algebraic curves of degree d states that $g(\sum_d) \geq \frac{1}{2}d^2 - \frac{3}{2} + 1$. The replacement of $\frac{11}{32}$ by $\frac{1}{2}$ in the lower bound for the genus, requires the substitution of 2 for $\frac{11}{8}$ as the lower limit for $\frac{b_2}{|\sigma|}$. Consider the intersection form $mE_8 \oplus nH$. Since

$$\begin{aligned} b_2(mE_8 \oplus nH) &= 8m + 2n \\ \sigma(mE_8 \oplus nH) &= 8m, \end{aligned} \tag{74}$$

the inequality

$$\frac{8m + 2n}{8m} \geq 2 \tag{75}$$

is equivalent to $n \geq 4\,m$. This very tight bound is not expected to be valid for a large class of smooth four-manifolds.

The $\frac{3}{2}$ conjecture for irreducible simply connected four-manifolds is $\chi \geq \frac{3}{2}|\sigma|$, where $\chi = 2 + rank(Q) = 2 + b_2$ is the Euler characteristic. It follows that

$$b_2 \geq \frac{3}{2}|\sigma| - 2. \tag{76}$$

Substituting the new coefficient into the lower bound for the genus of the embedded surface in an irreducible manifold,

$$g\left(\sum\right) \geq \left(\frac{3}{2} - \frac{2}{|\sigma|}\right)\left|\frac{2^r + 1}{3 \cdot 2^{r+1}}\sum^2 - \frac{2^{r-1}}{2^r - 1}\sigma(M)\right| - \frac{2^{r-1}}{2^r - 1}b_2(M), \tag{77}$$

where $2^r \| [\sum]$. Therefore, the genus of an algebraic curve of degree d embedded in an irreducible, simply connected manifold would satisfy the inequality $g(\sum_d) \geq \left(\frac{3}{8} - \frac{1}{2|\sigma|}\right)d^2 + \gamma_1 d + \gamma_0$ with γ_0 and γ_1 being constants.

7. Conclusion

The classification of four-manifolds has been reduced to the definite signature with odd intersection forms that are diagonal, $n\mathbf{1} + m(-\mathbf{1})$ or indefinite signature with even intersection forms $mE_8 + nH$, where E_8 is the exceptional Lie group Cartan matrix and H is the matrix $\begin{pmatrix} 0 & 1 \\ 1 & 0 \end{pmatrix}$, which is the form of $S^2 \times S^2$. All known smooth, oriented four-manifolds with an even intersection form are known to have coefficients satisfying $n \geq \frac{3}{2}|m|$ or equivalently, a second Betti number satisfying $b_2 \geq \frac{11}{8}|\sigma|$. The oriented, spin geometries in four dimensions have been demonstrated to admit a smooth structure only if $n \geq |m| + 1$ or $b_2 \geq \frac{5}{4}|\sigma| + 2$. The proof has been extended to non-spin manifolds with the inequality $n \geq |m| - \left(1 - \frac{1}{2^i}\right)$ and $b_2 \geq \frac{5}{4}|\sigma| - \left(2 - \frac{1}{2^{i-1}}\right)$. It is found here that the lower bound for the signed Betti number b_+ is larger than $2|k| + 1$, where $|k| = \frac{3}{16}|\sigma|$. Considering a cyclic covering of a non-spin manifold and introducing the degree into the proof for the spin manifold, the inequality $b_2 \geq \frac{11}{8}|\sigma|$. This increase in the lower bound for the second Betti number allows the existence of nonsmoothable manifolds with $b_2 \geq \frac{5}{4}|\sigma|$ within a strict demarcation between the regions for smooth and nonsmooth structures.

The existence of smooth, compact simply-connected manifolds with $\frac{5}{4} < \frac{b_2}{|\sigma|} < \frac{11}{8}$ present potential counterexamples to the $\frac{11}{8}$. The topological sums $\mathbb{CP}^2\#\overline{\mathbb{CP}^2}$ are included within these limits. It is proven in §4 that these spaces cannot have spin structures by Rohlin's theorem. Amongst the connected sums $\mathbb{CP}^2\#\ell\overline{\mathbb{CP}^2}$, only those values of ℓ congruent to 1 modulo 16 would satisfy the condition on the divisibility of the signature by 16. The absence of spin structures on $\mathbb{CP}^2\#(16r + 1)\overline{\mathbb{CP}^2}$ for $\ell \geq 1$ is established through the computation of the second Stiefel-Whitney class. Therefore, $\mathbb{CP}^2\#\overline{\mathbb{CP}^2}$ is the unique spin manifold in this set, which is necessary for consistency of the $\frac{10}{8}$ theorem. The conclusion on the nonexistence of oriented, compact simply connected four-manifolds, having both a smooth structure and a spin geometry, continues to be valid for topological sums of S^4, $S^2 \times S^2$ and $K3$ and complex algebraic surfaces, since the condition of the existence of a spin structure requires generally an increased minimum value of $\frac{11}{8}$ for $\frac{b_2}{|\sigma|}$.

Classification:

MSC: 57N13; 57R19

Author details

Simon Davis
Research Foundation of Southern California, La Jolla, CA, USA

*Address all correspondence to: sbvdavis@outlook.com

References

[1] Matsumoto Y. On the bounding genus of homology 3-spheres. Journal of the Faculty of Science of the University of Tokyo. Section I, Mathematics. 1982;**29**: 287-318

[2] Manolescu C. On the intersection forms of spin four-manifolds with boundary. Mathematische Annalen. 2014;**359**:695-728

[3] Eguchi T, Gilkey PB, Hanson AJ. Gravitation, gauge theories and differential geometry. Physics Reports. 1980;**66**:213-293

[4] Furuta M. Monopole equation and $\frac{11}{8}$-conjecture. Mathematical Research Letters. 2001;**8**:279-291

[5] Hopkins M, Lin J, Shi X, Xu Z. Intersection forms of spin 4-manifolds and the pin(2)-Equivariant Mahowald invariant. Communications of the American Mathematical Society. 2022;**2**

[6] Furuta M, Kametani Y, Matsue H. Spin 4-manifolds with signature=-32. Mathematical Research Letters. 2001;**8**: 293-301

[7] Kikuchi K. Representing positive homology classes of $\mathbb{CP}^2\#2\overline{\mathbb{CP}^2}$ and $\mathbb{CP}^2\#3\overline{\mathbb{CP}^2}$. Proceedings of the American Mathematical Society. 1993;**117**:861-869

[8] Barth W, Hulek K, Peters C, Van den Ven A. Compact Complex Surfaces. Heidelberg: Springer; 2004

[9] Furuta M. Finite-dimensional approximations in geometry. In: Li T, editor. Proceedings of the International Congress on Mathematics 2002. Vol. III. Beijing: World Scientific Publishing, Singapore; August 20-28, 2002, 2002. pp. 395-403

[10] Lawson B. The Theory of Gauge Fields in Four Dimensions, CMBS Regional Conference in Mathematics. Providence: American Mathematical Society; 1985

[11] Minami N. G-join theorem - an Unbased G-Freudenthal theorem, Homotopy representations, and a Borsuk-Ulam theorem. Nagoya, Japan: Nagoya Institute of Technology

[12] Schmidt B. Ein Kriterium für die Existenz äquivarianter Abbildungen zwishcen reellen Darstellungssphären der Gruppe Pin(2). University of Bielefeld: Diplomarbeit; 1997

[13] Furuta M, Kametani Y. The Seiberg-Witten Equations and Equivariant e-Invariants. Tokyo, Japan: University of Tokyo; 2001

[14] Furuta M, Kametani Y, Matsue H, Minami N. Homotpy theoretical considerations of the Bauer-Furuta stable Seiberg-Witten invariants. Geom. & Topology Monographs. 2007;**10**: 155-166

[15] Freedman MH, Taylor L. Λ-splitting of 4-manifolds. Topology. 1977;**16**: 181-184

[16] Kervaire M, Milnor JW. On 2-spheres in 4-manifolds. Proceedings of the National Academy of Sciences of the United States of America. 1981;**47**:1861-1657

[17] Goto R. Unobstructed deformations of generalized complex structures induced by C∞ logarithmic Symplectic structures and logarithmic Poisson structures. In: Futaki A, Miyaoka R, Tang Z, Zhang W, editors. 10th China-Japan Conference 2014. Geometry and Topology of Manifolds. Tokyo: Springer; 2016. pp. 159-184

[18] Kirby RC. T^3_{Lie} and $\mathbb{CP}^2\#^9(-\mathbb{CP}^2)$, the Topology of 4-Manifolds. Vol. 1374. Heidelberg: Lecture Notes in Mathematics; Springer-Verlag; 1989

[19] Park J. Simply connected symplectic 4-manifolds with $b_2^+ = 1$ and $c_1^2 = 2$. Inventiones Mathematicae. 2005;**159**: 657-667

[20] Kotschick D. On manifolds homeomorphic to $\mathbb{CP}^2\#\overline{\mathbb{CP}^2}$. Inventiones Mathematicae. 1989;**95**:591-600

[21] Donaldson SK. Irrationality and the h-Cobordism conjecture. Journal of Differential Geometry. 1987;**26**:141-168

[22] Lee R, Li T-J. Intersection forms of non-spin four manifolds. Mathematische Annalen. 2001;**319**:311-318

[23] Bohr C. On the signature of even 4-manifolds. Mathematical Proceedings of the Cambridge Philosophical Society. 2002;**132**:453-469

[24] Nakamura N. $Pin^-(2)$-monopole equations and intersection forms with local coefficients of 4-manifolds. Mathematische Annalen. 2013;**357**: 915-939

[25] Kotschick D, Matic G. Embedded surfaces in four-manifolds, branched covers, and SO(3)-invariants. Mathematical Proceedings of the Cambridge Philosophical Society. 1995; **117**:275-286

Chapter 5

Geometric Properties of Classical Yang-Mills Theory on Differentiable Manifolds

Paul Bracken

Abstract

Gauge theories make up a class of physical theories that attempt to describe the physics of particles at a fundamental level. The purpose here is to study Yang-Mills theory at the classical level in terms of the geometry of fiber bundles and differentiable manifolds. It is shown how fundamental particles of bosonic and fermionic nature can be described mathematically. The Lagrangian for the basic interactions is presented and then put together in a unified form. Finally, some basic theorems are proved for a Yang-Mills on compact four-dimensional manifolds.

Keywords: manifold, bundle, section, Yang-Mills, compact, four-dimensional, spinor, classical field

1. Introduction

In 1954 C. N. Yang and R. Mills proposed a classical field theory that incorporates Lie groups at a fundamental level [1]. Since then, great progress has been made in the area of subatomic physics by realizing that physics which is described by non-abelian Lie groups can display many novel features and play a major role in the kinds of physical theories they describe [2–7]. These features are alluded to having no classical analogu. When formulated using rigorous mathematics, Yang-Mills theories as well as gauge theories make elegant use of complicated structures called fiber bundles and associated vector bundles. These are indispensable in physics where spacetime, the base manifold has a non-trivial topology. This occurs in string theory for example spacetime is usually assumed to be a product $\mathbb{M}^4 \times K$ of Minkowski spacetime with a compact Riemannian manifold. If Euclidean spacetime $\mathbb{R}^4$ is compactified to the 4-sphere S^4, a similar situation applies [8–14]. Fields in spacetime often cannot be described simply by a map to a fixed vector space but as sections of a non-trivial vector bundle. In these cases, fields on spacetime often cannot be described simply by a map to a fixed vector space, but rather as sections of a nontrivial vector bundle [15–20].

The Lagrangian and action of a field theory should be invariant under the action of certain symmetry groups such as the Lorentz group, gauge symmetry, and conformal symmetry [21–26]. This means the Lagrangian for the fields, hence the laws of physics, are invariant under symmetry transformations. For spontaneously broken gauge theories, the Lagrangian is invariant under gauge transformations that originate in a Lie group. The Higgs condensate yields a vacuum configuration invariant only under a subgroup of G, and at the same time provides a mechanism for giving mass to particles. Although the quantum versions of theories are not discussed here, it is important to state that symmetries of the classical theory, such as gauge symmetries, do not necessarily hold in the quantized theory. The main reason is the quantization method, such as the path integral measure may not be invariant under the symmetry [27–31]. In this event, the theory is said to be anomalous.

In mathematical terms, suppose $\pi : E \to M$ is a surjective differentiable map between smooth manifolds. If $x \in M$ is an arbitrary point, the nonempty subset $E_x = \pi^{-1}(x) \subset E$ is called the fiber of π over x. For a subset $U \subset M$ we set $E_U = \pi^{-1}(U) \subset E$, the part of E above U, and it is the union of all fibers E_x, where $x \in U$. A differentiable map $s : M \to E$ such that $\pi \circ s = I_M$ where I_M is the identity map is called a global section of π. A differentiable map $s : U \to E$, defined on some open subset $U \subset M$ satisfying $\pi \circ s = I_U$ is called a local section. A differentiable map $s : U \to E$ is a local section of $\pi : E \to U$ if and only if $s(x) \in E_x$, for all $x \in U$. Fiber bundles are an important generalization of products $E = M \times F$ and can be understood as twisted products. The fibers are still embedded submanifolds and are all diffeomorphic. The fibration in general is only locally trivial, so locally a product which is not global.

Definition 1.1 Let, E, F, M be manifolds and $\pi : E \to M$ a surjective differentiable map. Then, $(E, \pi, M; F)$ is called a fiber bundle if: for every $x \in M$, there exists an open $U \subset M$ around x such that π restricted to E_U can be trivialized, so there is a diffeomorphism $\phi_U : E_U \to U \times F$ such that $\text{pr}_1 \circ \phi_U = \pi$. Denote a fiber bundle as $F \to E \to M$, E is called the total space, M the base manifold, F the general fiber, π the projection and (U, ϕ_U) a bundle chart.

Using a bundle chart, (U, ϕ_U), the fiber $E_x = \pi^{-1}(x)$ is seen to be an embedded submanifold of the total space E for every $x \in M$, and $\phi_{U_2} = \text{pr}_2 \circ \phi_U|_{E_x} : E_x \to F$ is a diffeomorphism between the fiber over $x \in U$ and the general fiber. For physical reasons, it is essential to include pseudo-Riemannian metrics in the picture. Let M be a smooth manifold. A pseudo-Riemannian metric g of signature (s, t) where $(+, \cdots +, -, \cdots, -)$ is a section that defines at each $x \in M$ a non-degenerate symmetric bilinear form $g_x : T_xM \times T_xM \to \mathbb{R}$ of signature (s, t).

Principal fiber bundles are a combination of the concepts of fiber bundle and group action; that is, fiber bundles have a Lie group action such that both structures can be made compatible. Let $G \to P \to M$ be a fiber bundle with general fiber a Lie group G and a smooth action $P \times G \to P$ on the right. For a principal G-bundle, the action of G preserves the fibers of π and is simply transitive on them. The orbit map $G \to P$ such that $g \to p \cdot g$ is a bijection for all $x \in M$, $p \in P_x$. There exists a bundle atlas of G-equivariant bundle charts $\phi_i : P_{U_i} \to U_i \times G$ satisfying $\phi_i(p \cdot g) = \phi_i(p) \cdot g$, for all $p \in P_{U_i}, g \in G$, where on the right G acts on pairs $(a, x) \in U_i \times G$ by $(x, a) \cdot g = (x, ag)$. The group G is called the structure group P. Two features distinguish a principal bundle $P \to M$ from a standard fiber bundle whose general fiber is a Lie group G: there exists a right G-action on P simply transitive on each fiber P_x, $x \in M$ and bundle P has a principal bundle atlas. If $P \to M$ is a principal G-bundle, $p \in P, g \in G$, τ_g denotes the right translation $p \to p \cdot g$. The fiber P_x is a submanifold of the total space P for every $x \in M$ and the orbit map $g \to p \cdot g$ is an embedding for all $p \in P_x$.

A fiber bundle $V \to E \to M$ is called a real or complex vector bundle of rank m if: The general fiber V and every fiber E_x for $x \in M$, are m-dimensional vector space over $\mathbb{K} = \mathbb{R}$ or $\mathbb{C}$, and there exists a bundle atlas $\{(U_i, \phi_i)\}_{i\in I}$ for E such that induced maps $\phi_{ix} : E_x \to V$ are vector space isomorphisms for all $x \in U_i$. Such an atlas is called a vector bundle atlas for E, and the chart a vector bundle chart. There are two features that distinguish a vector bundle $E \to M$ from a standard fiber bundle; the vector space structure on each fiber E_x, $x \in M$ and the bundle E has a vector bundle atlas. An example of this is the tangent bundle of a smooth manifold which is canonically a smooth real vector bundle [32–33].

Definition 1.2 Let G be a Lie group and M a Manifold. Suppose that $M \times G \to M$ is a right action. For $X \in g_L$ we define the associated fundamental vector field $\tilde{X}$ on M by

$$\tilde{X}_p = \frac{d}{dt}\bigg|_{t=0} (p \cdot \exp(tX). \tag{1}$$

If we denote by ϕ_p, the orbit map for the right action, $\phi_p : G \to M, g \to p \cdot g$, then

$$\tilde{X}_p = \left(D_e\phi_p\right)(X_p). \tag{2}$$

Similarly, suppose that $G \times M \to M$ is a left action. Then we define the fundamental vector field by

$$\tilde{X}_p = \frac{d}{dt}\bigg|_{t=0} (\exp(-tX) \cdot p), \tag{3}$$

for $p \in M$. If we denote by $\phi_{p'}$ the following orbit map for the left action, $\phi'_p : G \to M, g \to g^{-1} \cdot p$, then

$$\tilde{X}_p = \left(D_e\phi'_p\right)(X_e). \tag{4}$$

The fundamental vector field will also be denoted X_f when the presentation requires.

It is shown here that a physical theory can be constructed based on the idea of a differentiable manifold along with many other associated mathematical structures that can be defined on it. The result is a theory which can be used to describe fundamental interactions of elementary particles at the classical level. This also permits the introduction of other ideas which can have a physical influence such as topological invariants. There is no discussion with regard to quantization of gauge theories. These interactions include the strong and weak forces. Physically, Yang-Mills fields represent forces or carriers of force. The first half of the paper introduces most of the mathematical concepts needed to describe particles of both fermionic and bosonic nature. The last part specializes to Yang-Mills in four dimensions. It is discussed how the field equations can be obtained from a variational principle and how the theory of partial differential equations plays a role in their study.

2. Matter fields and couplings

Lie groups appear in principal bundles in gauge theories. These are associated to vector bundles which describe particles and where representations on vector spaces are built into gauge theories. Connections are associated with gauge fields and give rise to covariant derivatives representing interactions.

Definition 2.1 A connection one-form on a principal G bundle $\pi : P \rightarrow M$ is a one-form $A \in \Omega^1(P, g)$ on the total space P so that $r_g^* A = Ad_{g^{-1}} A$ for all $g \in G$ and $A(\tilde{X}) = X$ for all $X \in g_L$, where $\tilde{X}$ is the fundamental vector field associated to X and g_L the Lie algebra of G. This is called a gauge field in physics. □.

At $p \in P$, a connection one-form is a linear map $A_p : T_p P \rightarrow g_L$ and $Ad_{g^{-1}}$ is a linear isomorphism of the Lie algebra to itself. There is a correspondence between Ehresmann connections and connection one-forms. Physically we want certain objects to be gauge invariant. A global gauge transformation is a bundle automorphism of P or a diffeomorphism $f : P \rightarrow P$ which preserve the fibers of P and is G equivariant

$$\pi \circ f = \pi, \quad f(p \cdot g) = f(p) \cdot g, \quad p \in P, \quad g \in G. \tag{5}$$

Under composition of diffeomorphisms, the set of all gauge transformations forms a group $\mathrm{Aut}(P)$. A local gauge transformation is a bundle automorphism denoted $\mathrm{Aut}(P)$. In physics, gauge transformations are often defined as maps on the base manifold M to the structure group G even for non-abelian Lie groups.

Let $\pi : P \rightarrow M$ be a principal G-bundle. A physical gauge transformation is a smooth map $\pi : U \rightarrow G$ defined on an open set $U \subset M$. The set of all physical gauge transformations forms a group $C^\infty(U, G)$ with pointwise multiplication.

If $s : U \rightarrow P$ is a local gauge of the principal bundle on an open subset $U \subset M$, the local connection one-form or local gauge field $A_s \in \Omega^1(U, g_L)$ determined by s is defined as

$$A_s = A \circ Ds = s^* A. \tag{6}$$

Suppose we have a manifold chart on U and $\{\partial_\mu\}_{\mu=1,\cdots,n}$ are the local coordinate vector fields on U. Set $A_\mu = A_s(\partial_\mu)$ and choose a basis $\{e_a\}$ for the Lie algebra g_L and then expand A_μ over that basis

$$A_\mu = \sum_{a=1}^{\dim g_L} A_\mu^a e_a. \tag{7}$$

The corresponding real-valued fields $A_\mu^a \in C^\infty(U, \mathbb{R})$ and one forms A_s^a are called local gauge boson fields in physics.

A principal bundle can have many gauges and it is of interest to determine how the local connection one-forms transform as we change the local gauge. Let $s_i : U_i \rightarrow P$ and $s_j : U_j \rightarrow P$ be local gauges with $U_i \cap U_j \neq \emptyset$, then there exists $g_{ij}(x) : U_i \cap U_j \rightarrow G$ such that

$$s_{ij}(x) = s_j(x) \cdot g_{ji}(x), \qquad x \in U_i \cap U_j. \tag{8}$$

In (8), g_{ij} is the smooth transition function between local trivializations. There are local connection 1-forms $A_i \in A_{s_i} \in \Omega^1(U_i, g_L)$ and $A_j \in A_{s_j} \in \Omega^1(U_j, g_L)$ and it is desired to obtain the relationship between A_i and A_j. If $\mu_G \in \Omega^1(G, p)$ is the Mauer-Cartan form defined as $\mu_G(v) = D_g L_{g^{-1}}(v)$ for $v \in T_g G$, set $\mu_{ji} \in g_{ji}^* \mu_G \in \Omega^1(U_i \cap U_j)$. The theorem which follows accounts for the transformation of local gauge fields.

Theorem 2.1 The local connection one-form transforms as

$$A_i = Ad_{g^{-1}} \circ A_j + \mu_{ji} \tag{9}$$

on $U_i \cap U_j$. If $G \subset \mathbb{GL}(n, \mathbb{K})$ is a matrix Lie group then

$$A_i = g_{ji}^{-1} \cdot A_j \cdot g_{ji} + g_{ji}^{-1} \cdot dg_{ji} \tag{10}$$

where $\cdot$ denotes matrix multiplication, g_{ji}^{-1} the inverse of g_{ji} in G and dg_{ji} the differential of each component of the function $g_{ji} : U_i \cap U_j \to G \subset \mathbb{K}^{n \times n}$. If G is abelian, then $A_i = A_j + \mu_{ji} = A_j + g_{ji}^{-1} \cdot dg_{ji}$.

Proof: Let $s \in U_i \cap U_j$ and $Z \in T_x M$ and set

$$X = D_x s_j(Z) \in T_{s_j(x)} P, \qquad Y \in D_x g_{ji}(Z) \in T_{g_{ji}(x)} G. \tag{11}$$

with group action $\Phi : P \times G \to P$ given as $(p, g) \to pg$, we calculate using the differential of map $\Phi\ (X, Y) \to (D_X r_g)(X) + \mu_G(Y)_{xg}$, where r_g is right translation μ_G is the Mauer-Cartan form, and the chain rule

$$D_x s_i(Z) = D_x\left(\Phi \circ \left(s_j, g_{ji}\right)\right)(Z) = D_{s_j(x)} r_{g_{ji}}(X) + \mu_G(Y)_{s_j(x)}\Big|_f = D_{s_j(x)} r_{g_{ji}(x)}(X) + \mu_{ji}(Z)_{s_i(x)}\Big|_f. \tag{12}$$

By the defining properties of the connection form A, we have

$$\begin{aligned} A_i(Z) = A(D_x s_i(Z)) &= A\left(D_{s_j(x)} r_{g_{ji}(x)}(X) + \mu_{ji}(Z)_{s_i(x)}\Big|_f\right) = \left(r^*_{g_j i}(x) A\right)(X) + \mu_{ji}(Z) \\ &= Ad^{-1}_{g_{ji}(x)} \circ A_j(Z) + \mu_{ji}(Z). \end{aligned} \tag{13}$$

The second claim follows by recalling that for a matrix Lie group $Ad_{g^{-1}} \cdot a \cdot g$ for all $g \in G$, $a \in g_L$ and $\mu_G(v) = g^{-1} v$ for $v \in T_g G$

$$\mu_{ji}(Z) = \mu_G\left(D_x g_{ji}(Z)\right) = g_{ji}^{-1} \cdot dg_{ji}(Z). \tag{14}$$

Theorem 2.2 Let $P \to M$ be a principal bundle and $A \in \Omega^1(P, g_L)$ a connection one-form on P. Suppose that $f \in \mathcal{G}(P)$ is a global bundle isomorphism. Then $f^* A$ is a connection one-form on P

$$f^* A = Ad\,\sigma_{f^{-1}} \circ A + \sigma_f^* \mu_G. \tag{15}$$

Proof: This follows from the definition of a connection 1-form and the previous Theorem.

Let H be the associated horizontal vector bundle defined as the kernel of A. Then $TP = V \oplus H$ and we set $\pi_H : TP \to H$ for the projection onto the horizontal vector bundle.

Definition 2.2 The two-form $F \in \Omega^2(P, g_L)$ defined by

$$F(X,Y) = dA\,(\pi_H(X), \pi_H(Y)), \quad X, Y \in T_pP, \quad p \in P \tag{16}$$

is called the curvature two form of A. Sometimes F^A is written to emphasize the dependence on A. □

Definition 2.3 Let P be a manifold and g_L a Lie algebra. For $\eta \in \Omega^1(P, g_L)$ and $\phi \in \Omega^1(P, g_L)$, define $[\eta, \phi] \in \Omega^{k+l}(P, g_L)$ to be

$$[\eta, \phi](X_1, \ldots, X_{k+l}) = \frac{1}{k!l!} \sum_{\sigma \in S_{k+l}} \operatorname{sgn}(\sigma) \left[\eta\left(X_{\sigma(1)}, \ldots, X_{\sigma(k)}\right), \phi(X_{k+1}, \ldots, X_n)\right], \tag{17}$$

where the commutators on the right are the commutator in the Lie algebra g_L. This is often written $\eta \wedge \phi$ as well.

It is useful to recall that if $X = \tilde{V}$ be a fundamental vector field and Y a horizontal vector field on P, then the commutator $[X, Y]$ is horizontal.

Theorem 2.3 (Structure Equations) The curvature form F of a connection form A satisfies

$$F = dA + \frac{1}{2}[A, A]. \tag{18}$$

Proof: Eq. (18) can be checked by inserting $X, Y \in T_pP$ on both sides. Suppose X, Y are both vertical. Then X, Y are fundamental vectors $X = \tilde{V}_p$, $Y = \tilde{W}_p$ forcertain elements $V, W \in G_L$,

$$F(X,Y) = dA(\pi_H(X), \pi_H(Y)) = 0, \quad \frac{1}{2}[A,A](X,Y) = [A(X), A(Y)] = [V, W]. \tag{19}$$

The differential of a one-form A is given by

$$dA(X,Y) = L_X(A(Y)) - L_Y(A(X)) - A([X,Y]), \tag{20}$$

where vectors X, Y are extended to vector fields in a neighborhood of p. If the extension is chosen by the fundamental vector fields $\tilde{V}$ and $\tilde{W}$, then $dA(X,Y) = L_X(W) - L_Y(V) - [V, W] = -[V, W]$, since V, W are constant maps from P to g_L.

If both X and Y are horizontal $F(X,Y) = dA(X,Y)$ and $\frac{1}{2}[A,A](X,Y) = [A(X), A(Y)] = 0$.

If X is vertical and Y is horizontal, then $X = \tilde{V}_p$ for some $V \in g_L$, and we have

$$F(X,Y) = dA(\pi_H(X), \pi_H(Y)) = dA(0, Y) = 0, \quad \frac{1}{2}[A,A](X,Y) = [A(X), A(Y)] = [V, 0] = 0. \tag{21}$$

Thus since $\tilde{V}, Y]$ is horizontal

$$dA(X,Y) = L_{\tilde{V}}(A(Y)) - L_Y(V) - A\left(\left[\tilde{V}, Y\right]\right) = -A\left(\left[\tilde{V}, Y\right]\right) = 0. \tag{22}$$

Connections define an important idea in geometry: that of parallel transport in principal and associated vector bundles and leads to the concept of covariant derivative on an associated vector bundle. An interesting result is that if $X = \tilde{V}$ be a fundamental vector field and Y a horizontal vector field on P, then the commutator $[X, Y]$ is horizontal. In a similar way, F can be written locally as was done for the local section. If we have a manifold chart on U and $\{\partial_i\}$ are local coordinate basis vector fields on U, then $F_{\mu\nu} = F_s(\partial_\mu, \partial_\nu)$ and $F_{\mu\nu} = \sum_{a=1}^{\dim g_L} F^a_{\mu\nu} e_a$ and locally the structure equations take the form $F_{\mu\nu} = \partial_\mu A_\nu - \partial_\nu A_\mu + [A_\mu, A_\nu]$.

Definition 2.4 Let $\gamma : [a, b] \to M$ be a curve in M. The map

$$\Pi^A_g : P_{\gamma(a)} \to P_{\gamma(b)}, \quad p \to \gamma^*_p(b), \tag{23}$$

is called parallel transport in the principal bundle P along γ with respect to the connection A.

Similarly for a curve $\gamma : [0, 1] \to M$ the map $\Pi^{E,A}_\gamma : E_{\gamma(0)} \to E_{\gamma(1)}$, given by $[p, v] \to \left[\Pi^A_\gamma(p), v\right]$ is a well-defined and linear isomorphism, called parallel transport in the associated vector bundle E along the curve γ with respect to A. Let Φ be a section of E, $x \in M$ and $X \in T_xM$ a tangent vector. A covariant derivative is to be defined by choosing an arbitrary curve $\gamma : (-\varepsilon, \varepsilon) \to M$ with $\gamma(0) = x$, $\dot{\gamma}(0) = X$. For each $u \in (-\varepsilon, \varepsilon)$ parallel transport the vector $\Phi(\gamma(u)) \in E_{\gamma(u)}$ back to E_x along γ. Take the derivative at $u = 0$ of the curve which results in E_x giving an element in E_x. Formally, set

$$D(\Phi, \gamma, x, A) = \frac{d}{du}\bigg|_{u=0} \left(\Pi^{E,A}_{\gamma_u}\right)^{-1} (\Phi(\gamma(u)) \in E_x. \tag{24}$$

The restriction of the curve γ starting at 0 and ending at time u for $u \in (-\varepsilon, \varepsilon)$ is denoted γ_u. Parallel transport Π^A_γ is a smooth map between the fibers $P_{\gamma(a)}$ and $P_{\gamma(b)}$ and does not depend on the parametrization of the curve. Let γ be a curve in M from x to y and γ' a curve from y to z. Denote γ acting followed by γ' by $\gamma^*\gamma'$, where γ comes first, then $\Pi^A_{\gamma\circ\gamma'} = \Pi^A_{\gamma'}\circ\Pi^A_\gamma$.

Theorem 2.4 Let $s : U \to P$ be a local gauge $A_s = s^*A$ and $\phi : U \to V$ the map with $\Phi = [s, \phi]$. Then the vector $D(\Phi, \gamma, x, A) \in E_x$ is given by

$$D(\Phi, \gamma, x, A) = [s(x), d\phi(X) + \rho_*(A_s(X))\phi(x)]. \tag{25}$$

Proof: It holds that

$$\left(\Pi^{E,A}_{\gamma_t}\right)^{-1}(\Phi(\gamma(t))) = \left[\left(\Pi^A_{\gamma_t}\right)^{-1}(s(\gamma(t)), \phi(\gamma(H))\right]. \tag{26}$$

Let $q(t)$ be the unique smooth curve determined in the fiber P_x such that $\Pi^A_{\gamma(t)}(q(t)) = s(\gamma(t))$. Write $q(t) = s(x) \cdot g(t)$ and $g(t)$ is a uniquely determined smooth curve in G

$$\left(\Pi^{E,A}_{\gamma_i}\right)^{-1}(\Phi(\gamma(t))) = [q(t), \Phi(\gamma(t))] = [s(x), \rho(g(t))\phi(\gamma(t))]. \tag{27}$$

For $t = 0$, we have

$$s(x) = s(\gamma(0)) = \Pi^A_{\gamma_0}(q(0)) = q(0), \quad g(0) = a \in G. \tag{28}$$

Consequently, $\dot{g}(0) \in g_L$, and it follows that

$$D(\Phi,\gamma,x,A) = \frac{d}{dt}\bigg|_{t=0} [s(x), \rho(g(t))\phi(\gamma(t))] = [s(x), \rho_x(\dot{g}(0))\phi(x) + d\phi(X)]. \tag{29}$$

To finish, $\rho_r(\dot{g}(0))$ is calculated,

$$\frac{d}{dt}\bigg|_{t=0} s(\gamma(t)) = ds(X), \quad \frac{d}{dt}\bigg|_{t=0} \Pi^A_{\gamma_i}(g(t)) = \dot{q}(0) + \frac{d}{dt}\bigg|_{t=0} \Pi^A_{\gamma_i}(s(x)). \tag{30}$$

Since the curve $\Pi^A_{\gamma_t}(s(x))$ is horizontal, with respect to A, we obtain $A_s(x) = A(ds(X)) - A(\dot{q}(0))$. Since $\dot{q}(0)$ and $\dot{g}(0)_{s(x)}$ are related by ϕ_*, the map that associates to a Lie algebra element the corresponding vector field on M is a homeomorphism, hence, $A(\dot{q}(0)) = \dot{g}(0)$ by definition of connection one-form. It follows that

$$\rho_*(\dot{g}(0)) = \rho_*(A_s(X)), \tag{31}$$

and so the claim.

In fact, the theorem implies that $D(\Phi,\gamma,x,A)$ depends only on the tangent vector X not on the curve γ itself. Now we are in a position to define the covariant derivative.

Definition 2.5 Let Φ be a section of an associated vector bundle E and $X \in \mathcal{X}(M)$ a vector field on M. The covariant derivative $\nabla^A_X \Phi$ of the section of E defined by

$$\left(\nabla^A_X \Phi\right)(x) = D(\Phi,\gamma,x,A), \tag{32}$$

where γ is any vector through X_x tangent to γ. The covariant derivative is a map $\nabla^A : \Gamma(E) \to \Omega^1(M,E)$.

The fact that $\nabla^A \Phi$ is a smooth one-form in $\Omega^1(M,E)$ for every $\Phi \in \Gamma(E)$ is clear from the local formula. In physics the covariant derivative in a local gauge $s : U \to P$ with $\Phi = [s,\phi]$ is given as

$$\nabla^A_X \Phi = \left[s, \nabla^A_X \phi\right] \qquad \nabla^A_X \phi(x) = d\phi(X_x) + \rho_*(A_i(X_x))\phi(x). \tag{33}$$

The map ∇^A is $\mathbb{K}$-linear in both entries and satisfies $\nabla^A_{fX}\Phi = f\nabla^A_X \Phi$ for all smooth functions $f \in C^\infty(M,\mathbb{R})$. The Leibnitz rule $\nabla^A_X(\lambda\Phi) = (L_X\lambda)\Phi + \lambda\nabla^A_X\Phi$ holds for all smooth functions $\lambda \in C^\infty(M,\mathbb{K})$.

Suppose $\gamma : [0,1] \to M$ is a closed curve in M, $\gamma(0) = \gamma(1) = x$, a loop. Then parallel transport $\Pi^{E,A}_\gamma$ is a linear isomorphism of the fiber E_x to itself. This isomorphism is called the holonomy $Hol^E_{\gamma,x}$ of the loop γ in the basepoint x with respect to the connection A. The Wilson loop is the map W^E_γ that associates to a connection A and loop γ the number $W^E_\gamma(A) = \mathrm{Tr}\left(Hol^E_{\gamma,x}(A)\right)$.

The map ∇^A can be regarded as a generalization of the differential $d : C^\infty(M) \to \Omega^1(M)$. The differential d can be identified with the covariant derivative on the trivial line bundle over M. The differential can be uniquely be extended in the standard way to an exterior derivative $d : \Omega^k(M) \to \Omega^{k+1}(M)$ by demanding $ddf = 0$ for all $f \in C^\infty(M)$ and $d(\alpha \wedge \beta) = d\alpha \wedge \beta + (-1)^k \alpha \wedge d\beta$ for $\alpha \in \Omega^k(M)$ and $\beta \in \Omega^l(M)$. This differential satisfies $d \circ d = 0$ on all forms, and so the de Rham cohomology $H^k_{\mathbb{R}}(M)$ is well-defined for all k.

It is useful to show the covariant derivative can be extended similarly to an exterior covariant derivative

$$d_A : \Omega^k(M, E) \to \Omega^{k+1}(M, E). \tag{34}$$

This exterior covariant derivative, however, in general does not satisfy $d_A \circ d_A = 0$. There is a well-defined vector product $\wedge : \Omega^k(M) \times \Omega^1(M) \to \Omega^{k+1}(M, E)$ between standard differential forms, with values in $\mathbb{K}$, and differential forms with values in E. Here we get the product between a scalar in $\mathbb{K}$ and a vector in E, which is well-defined. Let ω be an element of $\Omega^k(M, E)$ and choose a local basis $e_1, \dots, e_r$ of E over an open set $U \subset M$, then ω can be written

$$\omega = \sum_{i=1}^{r} \omega_i \otimes e_i, \tag{35}$$

with uniquely defined k-forms $\omega_i \in \Omega^k(U)$.

It should be stated that the definition of forms can be extended by defining $C(M, W)$ as the set of all smooth maps from M into the vector space W, which has a canonical structure of a manifold, so that smooth maps are defined. A one-form on M with values in W is an alternating $C^\infty(M)$ nonlinear map $\omega : \chi(M) \times \cdots \times \chi(M) \to C^\infty(M, W)$. The set of all k-forms on M, and values in W can be identified with $\Omega^k(M, W) = \Omega^k(M) \otimes_{\mathbb{R}} W$. It is said forms in $\Omega^1(M, W)$ are twisted with W. Scalar product of twisted forms can be defined by choosing a local frame for E over $U \subset M$ and expand k-forms F, G twisted with E as $F = \sum_{i=1}^{r} F_i \otimes e_i$, $G = \sum_{i=1}^{r} G_i \otimes e_i$ with $F_i, G_i \in \Omega^k(U, \mathbb{R})$. Set

$$\langle F, G \rangle_E = \sum_{i,j=1}^{r} \langle F_i, G_j \rangle \langle e_i, e_j \rangle_E \tag{36}$$

with Hodge star operator $^* : \Omega^k(M, E) \to \Omega^{n-k}(M, E)$ by $^*F = \sum_{i=1}^{r} (*F_i) \otimes e_i$, and codifferential $d^* = (-1)^{t+nk+1*} d^*$.

Definition 2.6 Let ∇ be a covariant derivative on a vector bundle E. Define the exterior covariant derivative or differential $d_\nabla : \Omega^k(M, E) \to \Omega^{k+1}(M, E)$ by

$$d_\nabla \omega = \sum_{i=1}^{r} \left(d\omega_i \otimes e_i + (-1)^k \omega_i \wedge \nabla e_i \right). \tag{37}$$

If $\nabla = \nabla^A$ is the covariant derivative on an associated vector bundle determined by connection A on a principal bundle, write $d_A = d_\nabla$. □.

Theorem 2.5 The definition of d_∇ is independent of the choice of local basis $\{e_i\}$ for E.

Proof: Let $\{e_{i'}\}$ be another local basis of E over U. Then there exist unique functions $C_{ji} \in C(U, \mathbb{K})$ with $e'_i = \sum_{i=1}^{r} C_{ji} e_i$. The matrix C with entries C_{ji} is invertible. Let C^{-1} be the inverse matrix with entries C^{-1}_{ij} and define

$$\omega'_j = \sum_{l=1}^{r} C_{lj}^{-1} \omega_l. \tag{38}$$

Then $\omega = \sum_{i=1}^{r} \omega_i \otimes e_i = \sum_{j=1}^{r} \omega'_j \otimes e'_j$. Now let us calculate

$$\sum_{j=1}^{r} \left(d\omega'_j \otimes e'_j + (-1)^k \omega'_j \otimes \nabla e'_j \right) = \sum_{i,j,l=1}^{r} \left(d\left(C_{lj}^{-1} \right) \wedge C_{ji} \omega_l \otimes e_i + C_{lj}^{-1} C_{ji} d\omega_l \otimes e_i \right.$$
$$\left. + (-1)^k C_{lj}^{-1} \omega_l \wedge dC_{ji} \otimes e_i + (-1)^k C_{lj}^{-1} C_{ji} \omega_l \wedge \nabla e_i \right)$$
$$\sum_{i=1}^{r} \left(d\omega_i \otimes e_i + (-1)^k \omega_i \otimes e_i \right) + \sum_{i,j,l=1}^{r} \left(d\left(C_{lj}^{-1} C_{ji} + C_{ij}^{-1} dC_{ji} + C_{lj}^{-1} dC_{ji} \right) \wedge \omega_l \otimes e_i. \right.$$
$$\tag{39}$$

The last term is zero since

$$0 = d\delta_{li} = d\left(\sum_{j=1}^{r} C_{lj}^{-1} C_{ji} \right) = \sum_{j=1}^{r} \left(d\left(C_{lj}^{-1} \right) C_{ji} + C_{lj}^{-1} dC_{ji} \right). \tag{40}$$

The derivative d_∇ also satisfies

$$d_\nabla(\omega + \omega') = d_\nabla \omega + d_\nabla \omega', \qquad d_\nabla(\sigma \otimes e) = d\sigma + (-1)^k \sigma \wedge \nabla e, \tag{41}$$

as well as the Leibnitz formula for exterior covariant derivative. Unlike the case of the standard exterior derivative d, it can be shown that d_∇ in general has square $d_\nabla \circ d_\nabla \neq 0$, a fact related to the curvature F^∇ of the covariant derivative ∇.

3. Yang-Mills Lagrangians

In physics, the Lagrangians that are used are restricted out of an infinite set of possible Lagrangians by various principles. The Lagrangian or action of a field theory should be invariant under certain transformations of the fields by symmetry groups. The laws of physics have to be invariant as well, a second meaning of symmetry is invariance of the actual field configurations. In spontaneously broken gauge theories, the Lagrangian is invariant under gauge transformations with values in a given Lie group G. However, due to the Higgs field, the vacuum is invariant under a subgroup $H \subset G$ of transformations. The purpose of the Higgs is to give mass to the particles that appear in the Lagrangians without at the same time breaking gauge invariance. A quantum field theory associated to the Lagrangian should be renormalizable so after the renormalization of parameters, finite results that can be compared with experiment are obtained.

The scalar product of forms is given as

$$\langle \omega, \eta \rangle = \sum_{\mu_1 < \cdots < \mu_k} \omega_{\mu_1 \cdots \mu_k} \eta^{\mu_1 \cdots \mu_k} = \frac{1}{k!} \omega_{\mu_1 \cdots \mu_k} \eta^{\mu_1 \cdots \mu_k}, \qquad |\omega|^2 = \langle \omega, \omega \rangle. \tag{42}$$

To write the Yang-Mills equations, the Hodge star operator written as $^*\Omega^k(M, \mathbb{K}) \to \Omega^{n-k}(M, \mathbb{K})$ is the linear map on real-valued forms so that if dv_g is the volume element on M,

$$\omega \wedge * \eta = \langle \omega, \eta \rangle dv_g, \qquad \omega, \eta \in \Omega^k(M, \mathbb{R}). \tag{43}$$

The L^2-scalar product of forms $\langle \cdot, \cdot \rangle_{L^2} : \Omega_0^k(M, \mathbb{K}) \times \Omega_0^1(M, \mathbb{K}) \to \mathbb{K}$ is defined by

$$\langle \omega, \eta \rangle_{L^2} = \int_M \langle \omega, \eta \rangle dv_g. \tag{44}$$

To obtain a finite integral, it is usual to work with forms of compact support. The codifferential $d^* \Omega^{k+1} \to \Omega^k(M)$ is

$$d^* = (-1)^{t+nk+1*} d^*. \tag{45}$$

Theorem 3.1 Let M be a manifold without boundary. Then the codifferential d^* is the formal adjoint of the differential d with respect to the L^2 scalar product on forms of compact support $\langle d\omega, \eta \rangle_{L^2} = \langle \omega, d^* \eta \rangle_{L^2}$ for all $\omega \in \Omega_0^k(M)$, $\eta \in \Omega_0^{k+1}(M)$.

Proof: The difference $\langle d\omega, \eta \rangle - \langle \omega, d^* \eta \rangle$ with respect to the pointwise scalar product of the forms. Applying * twice gives a map $** : \Omega^{n-k}(M) \to \Omega^{n-k}(M)$ is given by

$$** = (-1)^{t+(n-k)k}. \tag{46}$$

Therefore, we have

$$\begin{aligned} \langle d\omega, \eta \rangle - \langle \omega, d^* \eta \rangle dv_g = (d\omega) \wedge^* \eta - \omega \wedge * (d^* \eta) = (d\omega) \wedge * \eta + (-1)^k \omega \wedge (d^* \eta) \\ = d(\omega \wedge * \eta). \end{aligned} \tag{47}$$

Stokes' Theorem applied here implies the result.

This knowledge allows us to define the covariant codifferential $d_\nabla^* : \Omega^{k+1}(M, E) \to \Omega^k(M, E)$ by

$$d_\nabla^* = (-1)^{t+nk+1*} d_\nabla{}^*. \tag{48}$$

To define the Yang-Mills Lagrangian and the associated Yang-Mills equations, procced as follows. To do so, we use an n-dimensional, oriented, psuedo-Riemannian manifold (M,g), with signature (s,t) a principal G-bundle $P \to M$ with compact structure group G of dimension r, a scalar product on g_L, which is Ad, invariant and an orthonormal vector space basis T_i for g_L.

Let A be a connection 1-form on the principal bundle P with curvature two-form $F^A \in \Omega^2(P, g_L)$. The curvature defines a twisted two-form $F_M^A \in \Omega^2(M, Ad(P))$. The Yang-Mills Lagrangian is defined by

$$\mathcal{L}_{YM} = -\frac{1}{2} \langle F_M^A, F_M^A \rangle_{Ad(P)}. \tag{49}$$

For a fixed connection A, this Lagrangian is a global smooth function $\mathcal{L}_{YM}(A) : M \to \mathbb{R}$. The Yang-Mills Lagrangian is gauge invariant, $\mathcal{L}_{YM}(f^* A) = \mathcal{L}_{YM}(A)$, for all bundle automorphisms $f \in \mathcal{G}(P)$ and all A on P. In a chart with coordinates x^μ, the components of F^A are $F_{\mu\nu}^A = F_s^A(\partial_\mu, \partial_\nu)$ and they can be expanded over the Lie algebra basis as

$$F^A_{\mu\nu} = F^{Aa}_{\mu\nu} T_a, \tag{50}$$

and $F^{Aa}_s \in \Omega^2(U)$ are real-valued differential forms, $F^{Aa}_{\mu\nu}$ are real-valued smooth functions on U. Thus, expanding (49), the Yang-Mills Lagrangian is locally

$$\mathcal{L}_{YM}(A) = -\frac{1}{2}\langle F^A_s, F^A_s \rangle = -\frac{1}{4} F^{Aa}_{\mu\nu} F^{A\mu\nu}_a, \tag{51}$$

where $F^{Aa}_{\mu\nu} = \partial_\mu A^a_\nu - \partial_\nu A^a_\mu + f_{bca} A^b_\mu A^c_\nu$, and structure constant f_{cba} for the Lie algebra.

Suppose (M,g) is compact and closed. The Yang-Mills action for a principal G-bundle $P \to M$ is the smooth map $S_{YM} : \mathcal{A}(P) \to \mathbb{R}$, with $\mathcal{A}(P)$ the space of all connection one-forms A on P defined by

$$S_{YM}(A) = -\frac{1}{2}\int_M \langle F^A_M, F^A_M \rangle_{Ad(P)} dv_g. \tag{52}$$

A connection A on the principal bundle P is a critical point of the Yang-Mills action if

$$\left.\frac{d}{du}\right|_{u=0} S_{YM}(A + u\beta) = 0, \tag{53}$$

for all such variations on P.

Theorem 3.2 A connection A on a principal bundle $P \to M$ is a critical point of the Yang-Mills action if and only if A satisfies the Yang-Mills equation

$$d_A {}^* F^A_M = 0. \tag{54}$$

Proof: Based on the structure equations, we calculate

$$F^{A+u\beta} = d(A+u\beta) + \frac{1}{2}[A+u\beta, A+u\beta] = F^A + u\,(d\beta + [A,\beta]) + \frac{1}{2}u^2[\beta,\beta]. \tag{55}$$

Differentiating this and using the adjoint property on M, it follows that

$$\left.\frac{d}{du}\right|_{u=0} \left\langle F^{A+u\beta}_M, F^{A+u\beta}_M \right\rangle_{Ad(P),L^2} = 2\langle d_A\beta, F^A_M \rangle_{Ad(P),L^2} = 2\langle \beta, d^*_A F^A_M \rangle_{Ad(P),L^2}. \tag{56}$$

The scalar product on the Lie algebra is non-degenerate, the L^2-scalar product is non-degenerate. It follows that A is a critical point of the Lagrangian (49) if and only if (54) holds.

Any connection A on P has to satisfy the Bianchi identity

$$d_A F^A_M = 0. \tag{57}$$

When the group $G = U(1)$, the local curvature forms are independent of the choice of local gauge s and define a global two-form F_A, so the Bianchi identity and Yang-Mills equations are given by $dF_M = 0$ and $d^* F_M = 0$. These are Maxwell's equations for a source-free electromagnetic field on a general n-dimensionsl oriented pseudo-Riemannian manifold.

Fields of different types can be introduced into the picture. These include matter fields that couple to the gauge field A, such as scalar fields or fermionic spinor fields, and are distinguished by the statistics they obey. These two types of particle are distinguished by an intrinsic property called spin, and this has to have its own treatment.

A complex scalar field is a smooth map $\phi : M \rightarrow \mathbb{C}$. A multiplet of complex scalar fields is a smooth map $\phi : M \rightarrow \mathbb{C}^r$ for some $r > 1$ with the standard Hermitian scalar product $\langle v, w \rangle = v^\dagger w$ on $\mathbb{C}^r$. Given a principal g-bundle $P \rightarrow M$ with compact structure group G of dimension r, a complex representation $\rho : G \rightarrow GL(W)$ with associated complex vector bundle E and G-invariant Hermitian scalar product $\langle \cdot, \cdot \rangle_W$ on W and bundle metric $\langle \cdot, \cdot \rangle_E$ on the vector bundle E. If the dimension of V is one, then a smooth section of E is called a multiplet of complex scalar fields and the vector space W is called a multiplet space. With the covariant derivative $d_A : \Gamma(E) \rightarrow \Omega^1(M, E)$ and the scalar product $\langle \cdot, \cdot \rangle_E$ on $\Omega^1(M, E)$, the Klein-Gordon Lagrangian can be given.

Definition 3.1 The Klein-Gordon Lagrangian for a multiplet of the complex scalar field $\Phi \in \Gamma(E)$ of mass m coupled to a gauge field A is

$$\mathcal{L}_{KG}(\Phi, A) = \langle d_A \Phi, d_A \Phi \rangle_E - m^2 \langle \Phi, \Phi \rangle_E. \tag{58}$$

For given fields Φ and A, the Klein-Gordon Lagrangian is a smooth function $\mathcal{L}_{KG}(\Phi, A) : M \rightarrow \mathbb{R}$.

The associated action $S_{KG}(\Phi, A)$ is the integral over the Klein-Gordon Lagrangian on the closed manifold M. In local coordinates on M, the kinetic term is

$$\langle d_A \Phi, d_A \Phi \rangle = -\langle \nabla^{A\mu} \Phi, \nabla^{A\mu} \Phi \rangle_E. \tag{59}$$

In a local gauge s for the principal bundle, the Klein-Gordon Lagrangian can be written as $\Phi|_U = [s, \phi]$

$$\mathcal{L}_{KG}(\Phi, A) = (\partial^\mu \phi)^\dagger (\partial_\mu \phi) - m^2 \phi^\dagger \phi + (\partial^\mu \phi)^\dagger (A_\mu \phi) - (\phi^\dagger A_\mu)(\partial^\mu \phi) - \phi^\dagger A^\mu A_\mu \phi. \tag{60}$$

As with the Yang-Mills Lagrangian, the Klein-Gordon Lagrangian of a multiplet of complex scalar fields coupled to a gauge field is gauge invariant.

To describe fermion fields classically using spinor fields on spacetime, a Lagrangian for fermions is defined. The setting for doing this is an n-dimensional oriented and time-oriented pseudo-Riemannian spin manifold (M,g) of signature (s,t), a spin structure $Spin^\dagger(M)$ together with complex spin bundle $S \rightarrow M$, and a Dirac form $\langle , \rangle$ on the Dirac spinor space, not necessarily positive definite, with Dirac bundle metric $\langle , \rangle_D$. We abbreviate $\langle \Psi, \Phi \rangle_D$ as $\overline{\Psi}\Phi$.

Definition 3.2 The Dirac Lagrangian for a free spinor field $\psi \in \Gamma(S \otimes E)$ mass m is defined by

$$\mathcal{L}_D(\psi) = \text{Re}\, \langle \Psi, D_A \Psi \rangle_{S \otimes E} - m \langle \Psi, \Psi \rangle_{S \otimes E} = \text{Re}\, \left(\overline{\Psi} D_A \Psi \right) - m \overline{\Psi} \Psi, \tag{61}$$

where $D_A \Gamma(S \otimes E) \rightarrow \Gamma(S \otimes E)$ denotes the twisted Dirac operator, the first term the kinetic term and the second pertains to the mass of the particle. The associated action $S_D[\Psi, A]$ is the integral over the Dirac Lagrangian on a closed manifold M.

Based on the fundamental Lagrangians which couple the fields to the gauge field, the Lagrangian of the Standard Model can be built up as the sum of all the individual Lagrangians that are to be accounted for and required to describe all the observed fields. It could be referred to as the Yang-Mills-Dirac-Higgs-Yukawa Lagrangian

$$\mathcal{L} = \text{Re}\left(\overline{\Psi} D_A \Psi\right) + \langle d_A \Phi, d_A \Phi \rangle - V(\Phi) - 2g_Y \,\text{Re}\left(\overline{\Psi}_L \Phi \Psi_R\right) - \frac{1}{2}\left\langle F_M^A, F_M^A \right\rangle_{Ad(P)}. \tag{62}$$

Experiment informs us that a realistic theory of particle physics has to involve chiral fermions with a nonzero mass because the weak interaction is not invariant under parity inversion.

4. Yang-Mills on four-dimensional manifolds

The general overview of Yang-Mills theory is now restricted to four-dimensional compact Riemannian manifolds. This will emphasize how Yang-Mills relates to manifolds which are the natural context for Yang-Mills theory for more than one reason. First the four-dimensional action is bounded below by the characteristic number of the bundle so the field is constrained by the topology. By invariant theory, this is linked to the conformal invariance of the action occuring just in dimension four. The base manifold conformal structure leads to the relevant geometry. The curvature is given in terms of the connection form ω, and the action is the sum of a gradient term and a non-linear self-interaction term. They are of comparable strength only in dimension four. Some of the symbols are adapted to the particular case studied here.

Riemannian geometry in dimension four is distinguished by the fact that the universal cover $Spin(4)$ of the rotation group $SO(4)$ is not a simple group, but factors $Spin(4) = SU(2) \times SU(2)$. One way to look at this is at the group level, $\mathbb{R}^4$ and $\mathbb{C}^2$ can be identified with the quaternions $\mathbb{H}$. Thus $SO(2)$ may be regarded as the unit quaternions. For unit quaternions, g and h, the map $x \rightarrow g^{-1}xh$ is an orthogonal transformation of $\mathbb{H} = \mathbb{R}^4$ with determinant one, and hence yields a homeomorphism $\pi : SU(2) \times SU(2) \rightarrow SO(4)$. This map has kernel $\{-1, 1\}$ and so indicates $SU(2) \times SU(2)$ as the two-fold universal covering group of $SO(4)$.

Suppose M is a Riemannian manifold, so the metric determines the basic Levi-Civita connection on the cotangent space

$$\nabla : \Gamma(T^*M) \rightarrow \Gamma(T^* \otimes T^*M). \tag{63}$$

Choosing a local basis of sections $\{e^i\}$ of T^*M we may write $\nabla e^i = \sum_k \omega_k^i \otimes e^k$, where $\{\omega_k^i\}$ are the connection one-forms. The nature of these one-forms can be understood in the context of an arbitrary bundle. Let G be a compact semi-simple Lie group with Lie algebra g_L and let $\pi : P \rightarrow M$ be a principal G-bundle over M. A connection on P is a choice of an equivariant horizontal subspace on T_*P or a g_L-valued one-form on P which has horizontal kernel and is equivariant $g^*\omega(X) = (Adg^{-1})\omega(X)$ for $x \in \Gamma(T_*P)$ and $g \in G$.

Let $\mathcal{C}$ denote the affine space of C^∞ connections of P. Then $\mathcal{C}$ becomes a vector space when a base connection is fixed. The equivariance property shows that the difference $\eta = \omega - \omega_0$ pulls down to M as a one-form with values in the adjoint bundle

$P\times_{Ad}g_L$ also denoted G_L. As such it determines a covariant map $\nabla : \Gamma(g_L) \to \Gamma(g_L)\otimes T^*(M)$ by virtue of $\phi \to \nabla_0 + [\eta, \phi]$, where ∇_0 is the covariant derivative corresponding to ω_0. If $\rho : G \to \mathrm{Aut}(\overline{E})$ is a representation and $E = P\times_\rho \overline{E}$ the associated vector bundle, then ω induces a covariant derivative

$$\nabla^E : \Gamma(E) \to \Gamma(\otimes T^*M) \tag{64}$$

on E by applying the Lie algebra representation $\rho : g \to End(\overline{E})$ to ∇ above. Suppose for example P is the frame bundle of T^*M, the Riemannian connection can be described either in terms of the covariant derivative or in terms of the corresponding $so(n)$-valued connection form $\omega = \{\omega_k^i\}$.

Given a connection ∇^E on a vector bundle E, several related operations can be constructed from ∇^E a the symbol map. Extending ∇^E to the covariant derivative $\overline{\nabla} = \nabla\otimes 1 + 1\otimes\nabla$ on $\Lambda^k\otimes E$, with ∇ the Riemann connection on Λ^*, and using exterior differentiation or its adjoint contraction as the symbol, an exterior differentiation $D : \Gamma(\Lambda^*\otimes E) \to \Gamma(\Lambda^{*+1}\otimes E)$ is obtained and its formal adjoint D^*. In a local orthonormal frame (e^i), $\phi\in\Gamma(\Lambda^*\otimes E)$,

$$D\phi = \sum_i e^i\wedge\overline{\nabla}_i\phi, \qquad D^*\phi = -\sum_i e_i\overline{\nabla}^i\phi. \tag{65}$$

There are also two second order operators. They are the trace Laplacian

$$\left(\nabla^E\right)^*\overline{\nabla}^E = -\sum_i \overline{\nabla}_i^E\overline{\nabla}_i^E - \overline{\nabla}^E_{\nabla_i e_i} \tag{66}$$

on $\Gamma(E)$, and the bundle Laplace-Beltrami operator $\square = (DD^* + D^*D)$ on $\Gamma(\Lambda^*\otimes E)$. The covariant derivative of $\nabla : \Gamma(g_L) \to \Gamma(g_L\otimes T^*M)$ extends by virtue of (65) to an exterior differentiation D on the space of sections $\Lambda^* = \Gamma(\Lambda^*\otimes g_L)$ by $D\phi = \nabla_0\phi + [\eta, \phi]$ where ∇_0 is the covariant derivative corresponding to ω_0.

The curvature of a connection ω on a principal bundle P is the g_L-valued two-form $\Omega(X, Y) = d\omega(hX, hY)$ where h is the projection onto the horizontal subspace of ω. One can say $D = d\cdot h$ is a derivation on equivariant g_L-valued one-forms on P given by $D\phi = d\phi + [\omega, \phi]$ for one-forms with vertical kernel and $D\phi = d\phi + \frac{1}{2}[\omega, \phi]$ for connection forms ϕ, in particular, $\Omega = d\omega + \frac{1}{2}[\omega, \omega]$ on P. Let us fix connection ω_0, so for any other ω, the difference $\eta = \omega - \omega_0$ descends to M as an element of A^1 and the difference of the curvature is

$$\begin{gathered}\Omega - \Omega_0 = d\eta + \frac{1}{2}[\omega, \omega] - \frac{1}{2}[\omega_0, \omega_0] = d\eta + \frac{1}{2}[\eta, \eta] + [\omega, \eta], \\ \Omega = \Omega_0 + D_0\eta + \frac{1}{2}[\eta, \eta].\end{gathered} \tag{67}$$

Alternatively, $\phi\in A^0$ lifts to an equivariant g_L-valued function on P and $D\phi$ has a vertical kernel, so

$$D\circ D(\phi) = d(D\phi) + [\omega, D\phi] = d(d\phi + [\omega, \phi]) + [\omega, D\phi] = [D\omega, \phi] - [\omega, D\phi] + [\omega, D\phi] = [D\omega, \phi]. \tag{68}$$

The formula descends to the base $D\circ\nabla(\phi)=[\Omega,\phi]$ for $\phi\in A^0$. In terms of a local basis of vector fields $\{e_i\}$ and dual forms $\{e^i\}$

$$\Omega(\phi)=D\circ\nabla\phi=D\left(\sum_i\nabla_i\phi e^i\right)=\sum_{j,k}\left(\nabla_j\phi e^k\wedge\nabla_k e^j+\nabla_k\nabla_j\phi e^k\wedge e^j\right). \tag{69}$$

For the Riemannian connection, $\nabla_i e^j-\nabla_j e^i=\left[e^i,e^j\right]$ so $\Omega_{ij}=\nabla_i\nabla_j-\nabla_j\nabla_i-\nabla_{[e_i,e_j]}$ and similarly for Ω^E.

On a four-dimensional Riemannian manifold, the metric covariant derivative on the spin bundle S_+ is a map $\nabla:\Gamma(S_+)\to\Gamma(S_+\otimes{}^*M)$. Thus ∇ on $S_\pm$ decomposes into two operators: first the Dirac operator $\mathcal{D}:\Gamma(S_\pm)\to\Gamma(S_\mp)$ where symbol is Clifford multiplication, and the twister operator $\overline{\mathcal{D}}:\Gamma(V_\pm)\to\Gamma(\Lambda^2_\pm)$, whose symbol is the orthogonal complement of Clifford multiplication. In a local orthonormal frame $\{e^i\}$ with $\phi\in\Gamma(V)$

$$\mathcal{D}\phi=\sum_i e^i\cdot\nabla_i\phi,\qquad \overline{\mathcal{D}}\phi=\nabla\phi+\frac{1}{4}\sum_i e^i\mathcal{D}\phi\otimes e^i. \tag{70}$$

The Dirac operator is elliptic and is formally self-adjoint on the total spin bundle S.

In four dimensions, the Riemannian curvature tensor $R\in\Lambda^2\otimes\Lambda^2$ decomposes under the splitting $\Lambda^2=\Lambda^2_+\oplus\Lambda^2_-$. Due to the symmetry $R_{ijkl}=R_{klij}$, this is an element of the symmetric tensor product $Sym^2\left(\Lambda^2_+\oplus\Lambda^2_-\right)$, which is a Spin (4) module, this breaks uo into five irreducible pieces.

The components of this tensor under this decomposition are $\left(W^+,\frac{R_s}{12},2B,W^-\right)$, where R_s is the scalar curvature, B the traceless Ricci tensor, and $W^\pm$ are the self-dual and anti-self-dual components of the conformally invariant Weyl tensor. This decomposition results in some important classes of four-manifolds: M^4 is Einstein if $B=0$, conformally flat if $W=0$, and self-dual (anti) if $W^-=0$ ($W^+=0$).

Suppose M is a spin four-manifold with Riemannian connection ∇ and E, a vector bundle over M with connection ∇^E and curvature Ω^E. Then the Dirac operator is $\mathcal{D}\Gamma(V\otimes E)\to\Gamma(V\otimes E)$ is defined for E-valued spinors by $\mathcal{D}=\sum_i e^i\cdot\overline{\nabla}_i$ where $\overline{\nabla}$ is the total covariant derivative on $V\otimes E$. It is shown this operator has an algebraic decomposition into Laplacian and curvature terms. Such an expression is called a Weitzenböck operator. These encompass more than one kind of operator so it is worth showing how they can be developed. To get $\mathcal{D}^2$, choose an orthonormal basis $\{e^i\}$ around $x\in M$, vector fields $\{e_i\}$ dual to the e^i such that $\left(\nabla_{e_i}e^j\right)_x=0$ for all i,j. Squaring $\mathcal{D}$ and separating the symmetric and skew-symmetric parts

$$\mathcal{D}^2=\left(\sum e^i\cdot\overline{\nabla}_i\right)\left(\sum e^j\overline{\nabla}_j\right)=\sum_{i,j}e^i\cdot e^j\cdot\overline{\nabla}_i\overline{\nabla}_j=-\sum_{i,j}\overline{\nabla}_j\overline{\nabla}_i+\sum_{i,j}e^i\cdot e^j\cdot\left(\overline{\nabla}_i\overline{\nabla}_j-\overline{\nabla}_j\overline{\nabla}_i\right). \tag{71}$$

This can be summarized as

$$\mathcal{D}^2=\overline{\nabla}^*\overline{\nabla}+\frac{1}{2}\sum_{i,j}e^i\cdot e^j\cdot R_{ij}\otimes 1+\frac{1}{2}\sum_{i,j}e^i\cdot e^j\cdot\left(1\otimes\Omega^E_{ij}\right). \tag{72}$$

Since ∇ is torsionless, $[e_i, e_j]_x = 0$, and the total curvature is $\Omega_{ij} = \overline{\nabla}_i \overline{\nabla}_j - \overline{\nabla}_j \overline{\nabla}_i$. The first term is (72) is the positive trace Laplacian of $\overline{\nabla}$, R_{ij} can be written in terms of the irreducible components of R.

Compact four-dimensional manifolds M^4 possess two real characteristic classes. These can be expressed locally as polynomials in the curvature of M and hence as polynomials in the irreducible components of the curvature $(R_s, B, W^{\pm})$.

Topological invariants arise in the consideration of four-dimensional manifolds M. These have two real characteristic classes. They are the Pontryagin class $p_1 M$ and the Euler class $\chi(M)$ given by

$$\begin{aligned} p_1 M &= \frac{1}{4\pi^2} \int_M \left(|W^+|^2 - |W^-|^2 \right) dv_g, \\ \chi(M) &= \frac{1}{8\pi^2} \int_M \left(\frac{R_s^2}{24} - 2|B|^2 + |W^+|^2 + |W^-|^2 \right) dv_g. \end{aligned} \tag{73}$$

More generally, if G is a compact simple Lie group, $H^i(BG; \mathbb{R})$ vanishes for $i = 1, 2, 3$ and is $\mathbb{R}$ for $i = 4$. Thus, there is a single real characteristic class for principal G-bundles over M^4 and it resides in dimension 4. In Yang-Mills theory, the corresponding characteristic number is called the Pontryagin index κ of the bundle. It is obtained by substituting the curvature Ω of P into the Killing form. In terms of the anti-self-dual components $\Omega^{\pm}$ of Ω, κ is

$$\kappa = \frac{1}{8\pi^2} \int_M \left(|\Omega^+|^2 - |\Omega^-|^2 \right) dv_g. \tag{74}$$

For functions on a bounded domain in $\mathbb{R}^n$ the Sobolov space $L_{k,p}(D)$ is the completion of the space of C^∞ functions in the norm

$$\|f\|_{k,p} = \left(\int_D \sum_{|\alpha|=1}^{k} |\partial_\alpha f|^p \right)^{1/p}. \tag{75}$$

These spaces are related by the Sobolev embedding theorems: for $p, q \geq 1$, the inclusion $K_{k,p}(D) \to L_{l.q}(D)$ is continuous for $k - n/p \geq 1 - n/q$, compact for $k - n/p > 1 - n/q$. This extends to vector bundles E with metric over M.

5. Deriving coupled Yang-Mills equations

It is the case that once the geometrical setting for a gauge theory has been set out, the requirement of naturality then determines the theory. Let $\pi : P \to M$ be a principal bundle over a four-dimensional M with compact simple structure group G, $\rho : G \to \mathrm{Aut}(\overline{E})$ a unitary representation of G, $E = p \times_\rho \overline{E}$ the associated vector bundle, and W the bundle associated to the frame bundle of M. To get a mathematically rigorous development of the field equations, assume that they are variational equations and arise as the stationary points of an action integral

$$A(g, \nabla, \phi) = \int_M L(g, \omega, \phi) dv_g, \tag{76}$$

where the Lagrangian is a 4-form constructed from g, ∇, and ϕ.

Then P is a manifold with a certain geometric structure. There is a free right action of G, so an automorphism of P is a map $f : P \to P$, which preserves this structure $f(xg^{-1}) = f(x)g^{-1}$ for all $x \in P$ and $g \in G$. Let $\mathrm{Aut}(P)$ denote the group of all bundle automorphisms f such that the induced map $\pi \cdot f : M \to M$ preserves orientation, $\mathrm{Aut}_0(P)$ the subgroup which induces the identity on M. In the language of physics, a section $s : M \ni U \to P$ is called a local choice of gauge, an automorphism $f \in \mathrm{Aut}_0(P)$ is a gauge transformation, and the group $\mathcal{G} = \mathrm{Aut}_0(P)$ is the gauge group of the bundle.

These properties of the Lagrangian are required (i) in a local coordinate system and choice of gauge, L should be a universal polynomial in $g, h, \Gamma, \phi, (\det g)^{-1/2}, (\det h)^{-1/2}$, and their derivatives, Γ the Christoffel symbols. (ii) the map L should be a natural transformation with respect to the bundle automorphism f, $L((\pi \cdot f)^* g, f^* \nabla, \rho(f^*)\phi) = f^* L(g, \nabla, \phi)$ (iii) it should have conformal invariance, for any σ on M, $L(e^{i\sigma} g, \nabla, \phi) = L(g, \nabla, \phi)$. Naturality with respect to $\mathrm{Aut}_0(P)$ means that $L(g, f^* \nabla, \rho(f^*)\phi) = L(g, \nabla, \phi)$. This is Weyl's principle of gauge invariance. For the case in which $L = L(g, \nabla)$ requiring naturality under orientation preserving diffeomorphisms of P, $SO(4)$ invariant theory implies

$$L = c_1|R_s|^2 + c_2|B|^2 + c_3|W^+|^2 + c_4|W^-|^2 + c_5\Omega\wedge\Omega + c_6\Omega\wedge *\,\Omega, \tag{77}$$

where $(R_s, B, W^{\pm})$ are the components of the Riemann curvature of g, Ω the curvature of ω and the c_i are real numbers. The actions of the various values of the c_i include topological invariants $p_1(M)$, $\chi(M)$ for example.

Let us be concerned with the action which depends on the bundle curvature which is called the Yang-Mills action

$$A(g, \nabla) = \int_M \Omega\wedge *\,\Omega dv_g = \int_M |\Omega|^2 \sqrt{\det(g)} dx^1 \wedge \cdots \wedge dx^4. \tag{78}$$

The action is evidently regular and Diff (M) covariant. It is conformally invariant because the $*$ operator on two-forms is

$$A(e^{2\sigma} g, V) = \int_M e^{-4\sigma} g^{ij} g^{kl} \langle \Omega_{ik}, \Omega_{jl} \rangle \left(\det(e^{2\sigma g})\right)^{1/2} dx^1 \wedge \cdots \wedge dx^4 = A(g, \omega). \tag{79}$$

A gauge transformation $g \in \mathcal{G}$ takes ∇ to $g\nabla g^{-1}$ and $\Omega = D \circ \nabla$ to $g\Omega g^{-1}$. The Lagrangian $|\Omega|^2$ is then unchanged because the Killing form is invariant. Since $|\Omega|^2 = |\Omega^-|^2 + |\Omega^+|^2$, (71) shows that $A(g, \omega) \geq 8\pi^2 k$ with equality if and only if $\Omega^- = 0$. Consequently, self-dual connections are absolute minima of the Yang-Mills action. There are two action integrals considered by physicists. They are the fermionic and the bosonic types.

Definition 5.1 The fermion action is defined on E-valued spinors $\psi \in \Gamma(V \otimes E)$ as

$$A(g, \nabla, \psi) = \int_M \left(|\Omega|^2 + \langle \psi, \mathcal{D}\psi \rangle \right) dv_g, \tag{80}$$

where $\mathcal{D}$ is the Dirac operator and $\langle , \rangle$ is the inner product on $V \otimes E$ and dv_g the volume form.

Definition 5.2 The boson action is defined on E-valued scalars $\phi \in \Gamma(E)$ by

$$A(g, \nabla, \phi) = \int_M \left(|\Omega|^2 + |\nabla \phi|^2 + \frac{5}{6}|\phi|^2 - V(\phi) \right) dv_g. \tag{81}$$

where $V : E \to \mathbb{R}$ is a gauge invariant polynomial on the fiber such that $\deg(V) \leq 4$.

Both Lagrangians are regular Diff (M) invariant and gauge invariant. The degree requirement comes about as we wish to vary the action over a Sobolev space and by the Sobolev inequality, any polynomial in ϕ whose degree does not exceed four is then integrable. Note the second term in the fermion Lagrangian is not positive definite, for suppose $\psi = \psi_+ - \psi_- \in \Gamma(Q_+ \oplus Q_-)$ satisfies $\mathcal{D}\psi = \lambda\psi$ for some eigenvalue λ, $\overline{\phi} = \phi_+ - \phi_-$ satisfies $\mathcal{D}\overline{\phi} = -\lambda\overline{\phi}$. This gives that the spectrum of $\mathcal{D}$ is symmetric about zero.

6. Theorems in four dimensions for the Yang-Mills system

Let us calculate the first variation of the action for a spinor field. Introduce two real parameters (u, v) and pick a one-parameter family of connections $\nabla_u = \nabla_0 + u\eta + \cdots$, $\eta \in \Gamma(\Lambda^1 \otimes g_L)$ and a one-parameter family of spinors $\psi_v = \psi_0 + v\psi + \cdots$, $\psi \in \Gamma(V \otimes E)$. The curvature and total covariant derivative on $V \otimes E$ are

$$\Omega_u = \Omega_0 + uD_0\eta + \frac{R_s}{2}[\eta, \eta], \qquad \overline{\nabla}_u = \overline{\nabla}_0 + u\rho(\eta). \tag{82}$$

Expanding the action, it is given by

$$A(\nabla_u, \psi_v) = \int_M \Bigg(|\Omega_0|^2 + 2u\langle \Omega_0 D_0 \eta\rangle + v\langle \psi \mathcal{D}\phi\rangle + v\langle \phi \mathcal{D}\psi\rangle \\ + u\left\langle \psi \sum_i e^i \rho(\eta_i)\psi \right\rangle + \cdots \Bigg) dv_g. \tag{83}$$

In (83), $\{e^i\}$ is a local orthonormal basis. This implies the equations, which result from the first variation are for $\eta \in \Gamma^\infty(\Lambda^1 \otimes g_L)$ and $\psi \in \Gamma^\infty(V \otimes E)$

$$\int_M \left(2\langle D^*\Omega, \eta\rangle + \left\langle \psi, \sum_i e^i \rho(\eta_i)\psi) \right\rangle \right) dv_g = 0, \qquad \int_M (\langle \psi, \mathcal{D}\psi\rangle + \langle \phi, \mathcal{D}\psi\rangle) dv_g = 0, \tag{84}$$

Recall that $\mathcal{D}$ is self adjoint, so (84) gives the pair of equations

$$D^*\Omega = J(\phi) = -\frac{1}{2}\sum_i \langle \psi, e^i \rho(\sigma^a)\psi\rangle \sigma_a \otimes e_i, \qquad \mathcal{D}\psi = 0. \tag{85}$$

In (85), $\{\sigma_a\}$ is a local orthogonal basis of sections of g_L, $\{\sigma^a\}$ the dual basis in $\Gamma(g_L^*)$. The current due to ψ is $J(\psi)$ and it is real-valued since $\langle \psi, e^i\rho(\eta_i)\psi\rangle = \langle e^i\rho(\eta_i)\psi, \psi\rangle$. It is interpreted as a one-form on the space of connections.

The boson action is defined on E-valued scalars $\phi \in \Gamma(E)$ by

$$A_b(g, \nabla, \phi) = \int_M \left(|\Omega|^2 + \left|\overline{\nabla}\phi\right|^2 + \frac{m}{6}|\phi|^2 - V(\phi) \right) dv_g. \tag{86}$$

The first variation of this action is computed as follows. Choose a one-parameter family of connections $\nabla_u = \nabla_0 + u\eta + \cdots$, $\eta \in \Gamma\left(\Lambda \otimes g_L\right)$ and a one-parameter family $\phi_v = \phi_0 + v\tau + \cdots$, $\tau \in \Gamma(V \otimes E)$

$$\Omega_u = \Omega_0 + uD_0\eta + \frac{u^2}{2}[\eta, \eta], \qquad \overline{\nabla}_u = \overline{\nabla}_0 + u\rho(\eta), \tag{87}$$

Hence, the action is

$$\begin{aligned} A(g, \nabla_u, \phi_v) = & \int_M |\Omega_0 + uD_0\eta|^2 + \left|\overline{\nabla}_0\phi_0\right|^2 + u\rho(\eta)\phi_0 + v|\nabla_0\tau|^2 + \frac{m}{6}|\phi_0 + v\tau|^2 \\ & + V(\phi_0 + v\tau)dv_g \end{aligned} \tag{88}$$

$$\begin{aligned} = & \int_M |\Omega_0|^2 + 2u\langle \Omega_0 D_0\eta\rangle + \left|\overline{\nabla}_0\phi_0\right|^2 + 2u\left\langle \overline{\nabla}_0\phi_0\rho(\eta)\phi_0\right\rangle \\ & + 2v\left\langle \overline{\nabla}_0\phi_0\overline{\nabla}_0\tau\right\rangle \\ & + \frac{m}{6}|\phi_0|^2 + \frac{m}{6}v\langle\phi, \tau\rangle + \frac{m}{6}v\langle\tau, \phi_0\rangle - V(\phi_0 + v\tau)\, dv_g. \end{aligned} \tag{89}$$

Differentiating with respect to u and v then setting $u = v = 0$,

$$\left.\frac{\partial A}{\partial u}\right|_0 = 2\int_M \left(\langle D_0^*\Omega_0, \eta\rangle + \left\langle \overline{\nabla}_0\phi, \rho(\eta)\phi_0\right\rangle\right) d\mu_g,$$

$$\left.\frac{\partial A}{\partial v}\right|_0 = \int_M \left(2\left\langle \overline{\nabla}_0\phi_0, \overline{\nabla}_0\tau\right\rangle + \frac{m}{6}\langle\phi_0, \tau\rangle + \frac{m}{6}\langle\tau, \phi_0\rangle + \langle V'(\phi_0), \tau\rangle \right) d\mu_g. \tag{90}$$

Equating the results, (90) to zero yields the coupled fermion and boson equations of motion taking $V' = a|\phi|^2\phi + m_b^2\phi$

$$\begin{aligned} & D^*\Omega = J = -\frac{1}{2}\sum \left\langle \phi, e^i \cdot \rho(\sigma^a)\phi\right\rangle \sigma_a \otimes e_i, && \mathcal{D}\phi = m\phi, \\ & D^*\Omega = J = -\operatorname{Re}\sum_i \left\langle \overline{\nabla}_i\phi, \rho(\sigma^a)\phi\right\rangle \sigma_a \otimes e_i, && \overline{\nabla}^*\overline{\nabla}\phi = \frac{m}{6}\phi + a|\phi|^2\phi + m_b^2\phi. \end{aligned} \tag{91}$$

In physics, one says Ω is a gauge field, ω its gauge potential, and ψ, ϕ the field of a massive particle interacting with Ω. When the fields are set equal to zero, the fermion and boson actions reduce to the Yang-Mills field equations. Self-dual connections satisfy this equation because they are absolute minima of the action. In fact, the first field equation can be used to get

$$D^*J = DD^*\Omega = [\Omega, {}^*\Omega] = *\sum_{\alpha,\beta} \left\langle \Omega_\alpha, \Omega_\beta\right\rangle \left[\sigma^\alpha, \sigma^\beta\right] = 0. \tag{92}$$

When the structure group is abelian, the equation $D^*J = 0$. This expresses the fact that electric charge is conserved in electromagnetism.

The field Eqs. (91) simplify considerably when we take $a = m = 0$. Then either $\mathcal{D}\psi = 0$ on E-valued spinors or $\overline{\nabla}^*\overline{\nabla}\phi = (m/6)\phi$ with ϕ an E-valued scalar.

Theorem 6.1 Let E be a vector bundle over a manifold M and (ϕ, Ω) a solution of the coupled boson equations $D^*\Omega = J$, $\overline{\nabla}^*\overline{\nabla}\phi = (R_s/6)\phi$. If M is a compact manifold with positive scalar curvature, or if $M = \mathbb{R}^4$ and ϕ vanishes at infinity, then $\phi = 0$ and Ω is Yang-Mills.

Proof: If M is compact and $s > 0$, integration by parts yields

$$\int_M \left(|\overline{\nabla}\phi|^2 + \frac{m}{6}|\phi|^2\right) dv_g = 0. \tag{93}$$

Thus $\phi = 0$ and $J = 0$. The equation $\overline{\nabla}^*\overline{\nabla}\phi = 0$ can be converted to a differential inequality for $|\phi|$

$$\begin{aligned} d^*d|\phi|^2 = 2d^*\langle\phi, \overline{\nabla}\phi\rangle = -2|\overline{\nabla}\phi|^2 + 2\langle\phi, \overline{\nabla}^*\overline{\nabla}\phi\rangle = -2|\overline{\nabla}\phi|^2, \\ d^*d|\phi|^2 = 2d^*(|d|\phi) = -2|d|\phi||^2 + 2|\phi|d^*d|\phi|. \end{aligned} \tag{94}$$

Thus upon solving the second in (94) for $|\phi|d^*d|\phi|$ and using the first, we get

$$|\phi|d^*d|\phi| = |d|\phi||^2 - |\overline{\nabla}\phi|^2 \leq 0. \tag{95}$$

Consequently, $\Delta|\phi| \geq 0$ If $|\phi|$ vanishes at infinity, the maximum principle implies that $\phi = 0$, hence the current J vanishes and Ω is a Yang-Mills field. $\square$.

Theorem 6.2 Let M be a compact Riemannian four-manifold with $R_s/3 - |W^-| \geq \varepsilon > 0$. There is a constant α such that (i) Any Yang-Mills Ω such that $\|\Omega^-\|_{0,2} < \alpha$ is self-dual (ii) Any solution (Ω, ϕ) to the massless coupled fermion Eqs. (90) with $\|\Omega^-\|_{0,2} < \alpha$ satisfies $\Omega^- = J = \phi^- = 0$.

Proof: (i) Start with the equations $D^*\Omega = D\Omega = 0$ to obtain $D\Omega^+ \pm D\Omega^- = 0$ so $D\Omega^- = D^+\Omega^- = 0$ and hence $\square\Omega^- = 0$. Here, $\square = DD^* + D^*D$ is the Laplace Beltrami operator. Integrate $\langle\Omega^-, \square\Omega^-\rangle$ by parts over M using the Weizenbock formula,

$$\square = \overline{\nabla}^*\overline{\nabla} + \frac{R_s}{3} + W^+() - \left[\left(\Omega^E\right)^+, \cdot\right] \tag{96}$$

and apply Kato's inequality gives

$$\begin{aligned} 0 = \langle\Omega^-, \square\Omega^-\rangle &= \int_M \left(\overline{\Omega}^- |\nabla^*|\nabla\Omega^- + \frac{m}{3}|\Omega^-|^2 + W^-|\Omega^-|^2 - \Omega^-\left[\left(\Omega^E\right)^-, \Omega^-\right]\right) dv_g \\ &= \int_M \left(|\overline{\nabla}\Omega^-|^2 + \left(\frac{m}{3} + W^-\right)|\Omega^-|^2 + \overline{\Omega}^-\left[\left(\Omega^E\right)^-, \Omega^-\right]\right) dv_g \\ &\geq \int_M \left(|d|\Omega^-||^2 + \varepsilon|\Omega^-|^2 - |\Omega^-|^3\right) dv_g. \end{aligned} \tag{97}$$

By Hölder's inequality, followed by Sobolev's inequality, the last term is bounded by

$$\int_M |\Omega^-|^3 dv_g \leq \alpha\|\Omega^-\|_{0,2} \cdot \left(\|d\Omega^-\|_{0,2}^2 + \|\Omega^-\|_{0,2}^2\right). \tag{98}$$

This is dominated by the first two terms whenever $\|\Omega^{-}\|_{0,2}$ is sufficiently small. But this means the right-hand side is positive and a contradiction. The only way this can be is that $\Omega^{-} = 0$.

(ii) If $\phi = (\phi^{+}, \phi^{-}) \in V_{+} \oplus V_{-}$ satisfies $\mathcal{D}\psi = 0$, then $\langle \phi^{-}, \mathcal{D}^2 \psi^{-} \rangle = 0$. Using the Weitzenböck formula for the squared Dirac operator $\mathcal{D}^2$, we obtain

$$
\begin{aligned}
0 &= \int_M \left(|\nabla \phi^{-}|^2 + \frac{R_s}{2} |\phi^{-}|^2 + \frac{1}{2} \sum_{i,j} e^i e^j \cdot \overline{\phi}^{-} \Omega_{ij}^{E} \phi^{-} \right) dv_g \\
&\geq \int_M |d|\phi^{-}||^2 + \frac{R_s}{2} |\phi^{-}|^2 + \frac{1}{2} \sum_{i,j} e^i e^j \overline{\phi}^{-} \Omega_{ij} \phi^{-} \, dv_g.
\end{aligned} \tag{99}
$$

Whenever $\|\Omega^{-}\|_{0,2}$ is sufficiently small this inequality can apply, provided that $\phi\phi^{-} = 0$ so then

$$
-2D^* \Omega = -2J = \langle \phi^{-}, e^i \rho(\sigma^{\alpha}) \phi \rangle \sigma_{\alpha} \otimes e_i + \langle \phi^{+}, e^i \rho(\sigma^{\alpha}) \phi^{-} \rangle \sigma_{\alpha} \otimes e_i = 0. \tag{100}
$$

Therefore, $\Omega^{-} = 0$ by (i).

Solutions of the coupled field equations have the properties expected of elliptic equations, specifically, for $p > 2$ an L_{2p} weak solution is C^{∞}. This is basically elliptic regularity. There is a subtle point in that the coupled equations are elliptic only after a choice of gauge. Rather than using a connection to identify $\mathcal{C} = A^1$, we shall choose a point $x \in M$ and ball $B = B(x; r)$ around x and fix a gauge, considered as a section of the frame bundle of E, to pull down connections. This identifies the space of connections over B with $A^1|_B$. Let V_0 be the connection corresponding to $0 \in A^1|_B$ under this identification. Then in terms of covariant derivatives, the original connection is $\nabla = d + \omega$, and V_0 is simply exterior differentiation d.

The tangent space to the orbit of the gauge group through $(\nabla, \phi) \in \mathcal{C} \in \mathcal{E}$ is the image of $K : A^0 \to A^1 + \mathcal{E}$ by $X \to (\nabla X, \rho(X), \phi)$. The L^2 orthogonal complement of the image, which is the kernel of the adjoint operator K^*, provides a natural slice for the gauge orbit. This adjoint is $K^* : (\eta, \psi) \to \nabla^* \eta + \langle \psi, \rho() \phi \rangle$, where this last term selects an element of $A^0 = (A^0)^*$ via the Killing metric. There is a theorem which applies at the regular points of $\mathcal{C} \times \mathcal{E}$, where the action of the gauge algebra is free which is just stated: Suppose M is a compact Riemannian 4-manifold possibly with boundary. If a regular field $(V, \phi) \in (\mathcal{C} \times \mathcal{E})_{k+1,p}$ with $k \geq 0$, $2 < p < 4$. Then there is a constant c such that for every field (η, ψ) with $\|(\eta, \psi)\|_{k+1,p} < c$ there is a gauge transformation $g \in \mathcal{C}_{k+2,p}$ unique is a neighborhood of the identity, with $K_{\lambda}^* (g \cdot (V + \eta) - V, g \cdot (\phi + \psi) - \phi) = 0$ weakly. If ∇, ϕ, η and ψ are C^{∞}, then g is C^{∞}.

Theorem 6.3 Let ∇ be an $L_{k+1,p}$, $k \geq 0$, $2 < p < 4$ connection on a bundle E over a four-manifold M and let $\sigma : M \to \text{Frame}(E)$ be a C^{∞} gauge for E. Then there exists a constant $c > 0$ depending only on M such that if $\nabla = d + \omega$ and $\|\omega\|_{k+1,p} < c$ in the gauge σ, then there is a gauge transformation $g \in \mathcal{G}_{k+2,p}$ such that $d^* \omega = 0$ in the gauge $g \cdot \sigma$. If ∇ is C^{∞}, then g is C^{∞}.

We can choose a C^{∞} gauge around a given point x_0 and modify this to a gauge in which $d^* \omega = 0$ using Theorem 6.2. To achieve this, it is necessary to make the $L_{k,p}$ norm of the fields small.

Theorem 6.4 Let ∇ be an $L_{1,p}$, $2<p<4$ connection on a domain $D\subset M^4$. Then there is a C^∞ gauge σ and a gauge transformation $g\in\mathcal{G}_{2,p}$ such that, after a constant conformal change of metric, $d^*\omega=0$ in a neighborhood of $0\in D$ in the gauge $g\cdot\sigma$ and the new metric.

Proof: Given $\varepsilon>0$ choose a C^∞ gauge in a neighborhood of $0\in D$ and a small ball $B(1,\tau)$, $\tau>1$ around 0 with $\|\omega\|_{0,p}<\varepsilon$ is the required scale. Take $B(1,\tau^2)$ to the unit disk $B(\tau,1)$ by a conformal change of metric. Since $|\omega|^{2p}$ and $|\nabla\omega|^p=\left|\sum e^k\otimes\nabla_k\omega\right|^p$ have conformal weight $2p$, rescaling gives

$$\|\omega\|_{0,2p,B(\tau,1)}^{2p}=\tau^{2p-4}\|\omega\|_{0,2p,B(1,\tau^2)}^{p},\qquad \|\nabla\omega\|_{0,p,B(\tau,1)}^{p}=\tau^{2p-4}\|\nabla\omega\|_{0,p,B(1,\tau^2)}^{p}. \tag{101}$$

In the new metric, Hölder's inequality gives,

$$\|\omega\|_{1,p,B(\tau,1)}\leq\|\nabla\omega\|_{0,p,B(\tau,1)}+c\|\omega\|_{0,2p,B(\tau,1)}\leq\|\nabla\omega\|_{0,p,B(1,\tau^2)}+c\|\omega\|_{0,2p,B(1,\tau^2)}\leq(1+c)\varepsilon, \tag{102}$$

where c^{2p} is the volume of the unit ball in the rescaled metric, which is uniformly bounded in τ for $\tau<1$. When ε is sufficiently small, Theorem 6.3 applies.

Uhlenbeck has proved the much more difficult fact that the rescaling used here depends only on $\|\Omega\|_{0,p}$.

Theorem 6.5 Let $(\nabla\phi)\in(\mathcal{C}\times\mathcal{E})_{1,p}$, $p>2$ be a weak solution to the coupled Yang-Mills Eq. (90). Then there is an $L_{2,p}$ gauge in which (∇,ϕ) is C^∞.

Proof: Fix an $x\in M$, By Theorem 6.2, there is an $L_{2,p}$ gauge defined in a neighborhood of x such that $\nabla=d+\omega$ with $d^*\omega=0$ in this gauge. Expanding the field equations in this gauge, we have $J=D^*\Omega=d^*d\omega+\omega d\omega+(1/2)\omega[\omega,\omega]$. Hence $dd^*\omega=0$, so $d^*d\omega=\Box\omega=\nabla^*\nabla\omega+\text{Ric}(\omega)$ by the Weitzenböck formula, $\Box=\nabla^*\nabla+\text{Ric}+(1/2)\sum e^i\cdot e^j\cdot\Omega_{ij}^E$. A boson field then weakly satisfies

$$\begin{gathered}\Delta\omega-\text{Ric}(\omega)-\omega d\omega-\frac{1}{2}\omega[\omega,\omega]-\text{Re}\,\langle(\nabla+\omega)\phi,\rho()\phi\rangle=0,\\ \Delta\phi+2\omega\phi+\omega\omega(\phi)+\frac{R_s}{6}\phi+a|\phi|^2\phi+m^2\phi=0,\end{gathered} \tag{103}$$

where Δ is the metric Laplacian on functions. Applying $\mathcal{D}$ to $\mathcal{D}\phi=m\phi$ and using the Weizenböck formula (72) for the square of the Dirac operator on E-valued spinors, gives equations for the fermion fields. These are uniformly elliptic systems. Regularity follows by usual elliptic theory.

7. Conclusions

An extensive theory of Yang-Mills fields coupled to scalar and spinor fields on finite dimensional manifolds has been established. As well as differential geometric ideas, the appearence and systematic use of non-abelian Lie groups is also crucial and as such play a deep role in the study of elementary particles. The Yang-Mills fields represent forces or more accurately, they can be thought of as carriers of those fundamental forces. The presentation has been innovative and proofs have been given for all of the theorems that were introduced. It can also be looked at as a starting point for the study of other topics such as the existence of singularities or isolated singularities.

Author details

Paul Bracken
Department of Mathematics, University of Texas, Edinburg, TX, USA

*Address all correspondence to: paul.bracken@utrgv.edu

References

[1] Yang CN, Mills RL. Conservation of isotopic spin and isotopic gauge invariance. Physical Review. 1954;**96**: 191-195

[2] Bleecker D. Gauge Theory and Variational Princples. Reading, MA: Addison-Wesley; 1981

[3] Weinberg S. The Quantum Theory of Fields. Vol. I. Cambridge: Cambridge University Press; 1995

[4] Weinberg S. The Quantum Theory of Fields. Vol. II. Cambridge: Cambridge University Press; 1996

[5] Jackiw R. Quantum meaning of classical field theory. Reviews of Modern Physics. 1977;**49**(3):681-706

[6] Beggs E, Majid S. Quantum Riemannian Geometry. NY: Springer-Verlag; 2020

[7] de Azcarraya J, Izquierda. Lie Groups, Lie Algebras, Cohomology and some Applications in Physics. Cmabridge: Cambridge University Press; 1998

[8] Lee JM. Introduction to Smooth Manifolds. NY: Springer-Verlag; 2013

[9] Nakahara M. Geometry. Adam Hilger, Bristol: Topology and Physics; 1990

[10] Hamilton MJD. Mathematical Gauge Theory. NY: Springer-Verlag; 2017

[11] Berline N, Getzler E, Vergne M. Heat Kernels and Dirac Operators. Berlin: Springer-Verlag; 2004

[12] Atiyah MF, Bott R. The Yang-Mills equation over Riemann surfaces. Philosophical Transactions. Royal Society of London. 1983;**A 308**:523-615

[13] Schwinger J. Gauge invariance and Mass. Physical. Review. 1962;**125**: 397-398

[14] t'Hooft G. Renormalizable Lagrangians for massive Yang-Mills fields. Nuclear Physics. 1971;**B 351**: 167-188

[15] Parker T. Gauge theories on four dimensional Riemannian manifolds. Communications in Mathematical Physics. 1982;**85**:563-602

[16] Hitchin N. Linear field equations on self-dual spaces. Proceedings of the Royal Society of London. A. Mathematical and Physical Sciences. 1981;**A 370**:173-191

[17] Atiyah M, Bott R, Patodi V. On the heat equation and the index theorem. Inventiones. 1973;**19**:279-330

[18] Bracken P. Classical Yang-Mills field equations coupled to a scalar triplet invariant under subgroups of the conformal group. Journal of Geometry and Physics. 1999;**31**:229-240

[19] Bracken P. Invariant solutions of the Yang-Mills field equations coupled to a scalar doublet. Letters in Mathematical Physics. 1999;**47**:19-31

[20] Bracken P. Hamiltonian formulation of Yang-Mills theory and gauge ambiguity in quantization. Canada Journal of Physics. 2003;**81**: 545-554

[21] Abers E, Lee BW. Gauge theories. Physics Reports. 1972;**9**(1):1-141

[22] Goldstone J, Salam A, Weinberg S. Broken Symmetries. Physics Review Letters. 1962;**127**:956-970

[23] Weinberg S. Physical processes in a convergent theory of the weak and electromagnetic interaction. Physics Review Letters. 1971;**27**:1688-1691

[24] Ivanencko D, Sardanashvily G. The gauge treatment of gravity. Physics Reports. 1983;**94**(1):1-45

[25] Rennie R. Geometry and topology of chiral anomalies in gauge theories. Advances in Physics. 1998;**39**(6):617-779

[26] Leibbrandt G. Introduction to covariant gauges. Reviews of Modern Physics. 1987;**59**(4):1067-1119

[27] Aldrovandi R, Pereira JG. An Introduction to Geometrical Physics. Singapore: World Scientific; 2017

[28] Buonchotiani AM. An algebra of the Yang-Mills field. Journal of Mathematics Physics. 1973;**14**:849-854

[29] Balachandran AP, Marmo G, Skygerstam B, Stern A. Gauge Symmetries and Fibre Bundles. Lecture Notes in Physics. 1983;**188**, Springer-Verlag

[30] Birmingham D, Blau M, Rekowski M, Thompson G. Topological field theory. Physics Reports. 1991;**209**: 129-340

[31] Uhlenbeck K. Removable singularities in Yang-Mills fields. Communications in Mathematical Physics. 1982;**80**:11-30

[32] Vinet L. Matter-coupled Yang-Mills system in Minkowski space I: Invariant solutions in the presence of scalar fields. Physics Review. 1981;**D 24**:3179-3193

[33] Harnad J, Schneider S, Vinet L. The Yang-Mills system in compactified Minkowski space, invariance conditions and $SU(2)$ invariant solutions. Journal of Mathematical Physics. 1979;**20**:931-942